BIOTECHNOLOGY IN AGRICULTURE, INDUSTRY AND MEDICINE

BIOINFORMATICS RESEARCH

NEW DEVELOPMENTS

Biotechnology in Agriculture, Industry and Medicine

Additional books in this series can be found on Nova's website
under the Series tab.

Additional E-books in this series can be found on Nova's website
under the E-book tab.

Computer Science, Technology and Applications

Additional books in this series can be found on Nova's website
under the Series tab.

Additional E-books in this series can be found on Nova's website
under the E-book tab.

BIOTECHNOLOGY IN AGRICULTURE, INDUSTRY AND MEDICINE

BIOINFORMATICS RESEARCH

NEW DEVELOPMENTS

CHIHEB BATTIK
AND
KHALIL BELHASSINE
EDITORS

Nova Science Publishers, Inc.
New York

Library of Congress Cataloging-in-Publication Data

ISBN: 978-1-61942-363-3

Published by Nova Science Publishers, Inc. ✝ New York

Contents

Preface

This book presents new developments in the study of bioinformatics research. Topics discussed in this compilation include natural selection, molecular recombination structure and evolutionary historical relationships in bioinformatics; gene-gene and gene-environment causal interactions from different types of genomic data using bioinformatics; bioinformatics analysis of gene networks involved in genomic stability and cancer and protein bioinformatics for drug discovery.

Chapter I – Bioinformatics is a new field of research, with roots in computer science, physics, statistics and molecular biology. Its growth was fueled by the genome sequencing initiatives and the great amount of data produced by them, having quickly spread to all other fields of biological research. This innovative way of making science proved to be especially useful to immunology, allowing the development of a range of tools to predict critical steps in the immune response. The Immunological bioinformatics, or just Immunoinformatics, has emerged as a promising field, with possible applications in cancer research and vaccine development. Most of these immunoinformatics tools and databases are based on sequence analysis, which brings advantages, but also some limitations. Bioinformatics, however, has already developed ways to deal with tridimensional structures, allowing studying protein-protein interaction at an atomic level and extract, from each amino acid in these proteins, much more information than that provided by the linear sequence. In this chapter, we will discuss new uses of structural bioinformatics to study immunological problems, focusing on the Major Histocompatibility Complex (MHC). The MHC molecule plays a key role in the cellular immune response, presenting peptides derived from intracellular proteins to Cytotoxic T Lymphocytes (CTLs). Through the interaction between these peptide: MHC complexes (pMHCs) and the T Cell Receptors (TCRs), these CTLs are capable of identifying non-self peptides and eliminate infected or transformed cells. Structural bioinformatics tools, such as Molecular Docking and Molecular Dynamics, are being used to build the tridimensional structure of unknown pMHC complexes – not yet determined by classical experimental methods – and to assess molecular aspects of these structures. For instance, the electrostatic potential of the TCR-interacting surface of these pMHCs, as well as Accessible Surface Area (ASA) of the peptide inside the MHC cleft, can be related to the immunogenicity of a given virus-derived epitope. Moreover, these features can be included in one global analysis that can cover most of many details underlying TCR/pMHC interaction, affording to predict the impact of viral escape mutations and selection of epitopes with potential to induce cross-reactive immune responses.

Chapter II – In order to determine putative recombination events and their frequency in the RNAs 1 and 2 of currently the known members of the subfamily *Comovirinae*, a detailed study comprising 107 accessions retrieved from the international databases, was carried out by using RECCO and RDP v3.31β softwares. These programs allowed the detection of potential recombination sites in all the three virus genera composing this subfamily with various degrees of consistency. Furthermore, based on Fisher's Exact Test of Neutrality and contrary to RNA 1, positive selection exerted on protein-coding genes was detected in the RNA 2 of numerous representatives included in the subfamily *Comovirinae.* The reconstructed trees showed that *Raspberry ringspot virus* constitutes an independent cluster clearly separated from members of the current Subgroup A of *Nepovirus* genus. In addition, *Grapevine deformation* and *Grapevine Anatolian ringspot* viruses should be included in the subgroups A and B of the same genus, respectively, instead of being considered as two unclassified nepoviruses.

Chapter III – In this chapter we will review bioinformatics approaches in modeling the gene-gene and gene-environment causal interactions of systems biology studies, such as Genome-wide association studies (GWAS) and Microarray gene expression studies. Many gene-gene causal interactions models have been introduced recently. However, given many different types of genomic data – such as gene sequence, gene expression, etc., not many bioinformatics approaches are available in combining knowledge extracted from the data. To this end, we will review gene-gene and gene-environment interaction in terms of Single Nucleotide Polymorphism (SNP) and gene expression studies. We introduce a promising probabilistic causal model, i.e., causal Bayesian networks, and discuss how to represent gene-environment causal interactions using the probabilistic causal model. We conclude this chapter with promising research directions of gene-gene and gene-environment causal interactions models using bioinformatics approaches.

Chapter IV – Genetic variation of viruses is shaped by two major evolutionary forces: recombination and positive selection. In identifying putative recombination events in the complete genome of currently the known members of the family *Betaflexiviridae*, comprising 91 accessions retrieved from the international databases, in-depth molecular investigations were conducted by using RDP version 3.31β algorithm. It was demonstrated that potential recombination events were detected in three out of six virus genera (i.e, *Capillovirus*, *Foveavirus*, *Carlavirus*) along with the isolate TvAQ7 of the vitivirus *Grapevine virus E* and the unassigned *Betaflexiviridae*: *Sugarcane striate mosaic-associated virus*. Moreover, based on Fisher's Exact Test of Neutrality, positive selection operating on amino acid sequence levels of all virus genera except *Foveavirus* genus and the isolate DMV-932 of the citrivirus *Dweet mottle virus,* was also detected. It is worth noting that recombination and positive selection were both identified in representatives of *Capillovirus* and *Carlavirus* genera together with the unassigned *Betaflexiviridae*: *Sugarcane striate mosaic-associated virus*. The evolutionary history of all members of the family *Betaflexiviridae* showed that they segregated into seven distinct clusters corresponding to the six described virus genera and the clade "Unassigned *Betaflexiviridae*". From the present phylogenetic study, at least two taxonomic implications were drawn: (i) *Passiflora latent virus* should be considered as a typical carlavirus instead of belonging to the grouping "Unclassified carlaviruses". (ii) *Cherry virus A*, *Potato virus T*, *Grapevine virus E* and *Fig latent virus* should be extracted from the current groupings as actually stated by reports of the International Committee on Taxonomy of Viruses and gathered into a coherent ensemble to which the name "Unclassified *Betaflexiviridae*" should be allocated.

Chapter V – Positive selection and recombination are two major driving forces of virus evolution. In assessing potential recombination events and their frequency through the sequence of the whole genome of 73 accessions of the family *Alphaflexiviridae* retrieved from the international databases, a detailed study was conducted by using two algorithms: RECCO and RDP version 3.31β. These programs allowed the detection of potential recombination sites in all the six virus genera composing this family with various degrees of consistency. Furthermore, based on Fisher's Exact Test of Neutrality, no positive selection exerted on protein-coding genes was detected. Inferred phylogenies showed that putative recombination events played a major role in the promiscuous evolutionary history of the family *Alphaflexiviridae.* In fact, contrary to the statement of the International Committee on Taxonomy of Viruses, all currently the known isolates of *Potato virus X* were cleaved into two subgroups. While the smallest subgroup gathered nonrecombinant isolates, the largest encompassed recombinants. Similarly, members of *Potexvirus* genus split into three distinct groups. Although groups I and III comprised several recombinants, the group II was constituted by the two nonrecombinant isolates of *Strawberry mild yellow edge virus.* Moreover, all recombinant unclassified potexviruses were revealed to be phylogenetically closely related to recombinant typical potexviruses. As a result, a reclassification should be considered.

Chapter VI – In assessing putative recombination events and their frequencies in the complete genome of 168 accessions extracted from the international databases, in-depth molecular investigations were carried out by using two algorithms adapted to the mosaic structure of viruses: RECCO and RDP version 3.31β. By dissecting the recombination frequency through the sequences, it was demonstrated that the six genera of the family *Virgaviridae* encompass numerous members subjected to recombination. It was revealed that: (i) recombination events were widespread in the genome of several members of the six genera composing this family, (ii) recombination events were more frequent in members of *Furovirus, Pecluvirus, Pomovirus,* and *Tobravirus* genera than those of *Hordeivirus* and *Tobravirus* genera, (iii) in majority, swapped fragments had a length exceeding three nucleotides reaching a size of as long as 952 residues (*Tobacco mosaic virus*.152 isolate), (iv) as assessed by RDP package, heredity material exchange arose between non homologous sequences like those of *Indian peanut clump* and *Peanut clump pecluvirues.* Furthermore, based on Fisher's Exact test of Neutrality, positive selection operating of protein expression was detected only in *Beet soil-borne* and *Potato mop-top pomoviruses* as well as in the typical tobamovirus *Tomato mosaic virus* and the unclassified tobamovirus *Maracuja mosaic virus.* According to intra-genera analyses, inferred phylogenies showed that: : (i) within segmented viruses, each delineated cluster corresponded to a divided genomic region, (ii) *Tobacco mosaic virus* isolates retrieved into three groupings, (iii) *Tomato mosaic* and *Tobacco mosaic tobamoviruses* may have common ancestors, (iv) unclassified tombusvirus members did not constitute a phylogenetically homogenous group.

Chapter VII – Prion diseases cover a large range of neurodegenerative diseases in humans and animals, which are invariably fatal and highly infectious. By now there have not been some effective therapeutic approaches or medications to treat all prion diseases. Fortunately, numerous experimental experiences have showed that rabbits are resistant to infection from prion diseases isolated from other species, and recently the molecular structures of rabbit prion protein and its mutants were released into protein data bank. Prion diseases are "protein structural conformational" diseases. Thus, in order to reveal some secrets of prion diseases, it is amenable to study rabbits by techniques of the molecular structure and its dynamics. Wen et al. (PLoS One 5(10) e13273 (2010), Journal of Biological Chemistry 285(41) 31682-31693 (2010)) reported the surface of NMR $RaPrP^C$(124-228) molecular snapshot has a large land of continuous positive charge distribution, which contributes to the structural stability of rabbit prion protein. This is just the property at one snapshot. This paper confirms a large land of positive charge distribution has still been reserved during the long molecular dynamics of 30 nanoseconds in many environments, but the continuous of the land has not been always reserved yet. These results may be useful for the medicinal treatment of prion diseases.

Chapter VIII – Chromosomal abnormalities, somatic mutations and epigenetic alterations are among the major cellular changes associated withthe impairment of genome maintenance mechanisms (GMM) in solid tumors. Together with apoptosis, the GMM gene network is essentially characterized by the four major DNA repair mechanisms: nucleotide excision repair (NER), base excision repair (BER), mismatch repair (MMR) and recombinational repair (RER)—homologous recombinational repair (HRR) and nonhomologous endjoining (NHEJ). The GMM gene network is highly inter-connected. Its dysfunction contributes substantially to cancer genome instability, which hasimportant clinical implications, since it imposes technical problems for tumor biomarker development, such as the low recurrence of somatic mutations causally related to cancer and the extensive tumor heterogeneity. Using statistical, systems biology and computational approaches, in this chapter we consider different types of molecular changes that can imbalance the genome of cancer cells together with current methods used to study high-throughput data in the context of apoptosis and genome stability systems. In the introduction we discuss the molecular features that are commonly attributed to solid tumors, the theories that describe the tumor progression and the resulting DNA damage accumulation patterns. In section two we review some computational methods used to characterize gene expression networks, some available internet resources that can be used to study genome instability and the possible strategies to assess tumor biomarker networks. Finally, in section three, we present several applications illustrating the use of the computational methods previously defined, from the analysis of cancer-related gene networks using microarray data and gene knockout, to the analysis of tissue homeostasis and cell growth control in the perspective of possible evolutionary scenarios that gave rise to genome maintenance mechanisms in multicellular organisms.

Chapter IX – The field of protein bioinformatics analyzes the sequence and structure of protein; it plays a critical role in the discovery of small therapeutic agents as well as protein drugs. Here we present our recent progress in this field, including the new concepts of concavity druggability and antibody druggability, which are expected to raise the drug discovery success ratio.

In: Bioinformatics Research: New Developments
Editors: C. Battik and K. Belhassine
ISBN: 978-1-61942-363-3

Chapter I

Structural Immunoinformatics and Vaccine Development

D. A. Antunes, M. M. Rigo, M. Sinigaglia and G. F. Vieira
NBLI - Núcleo de Bioinformática do Laboratório de Imunogenética,
Department of Genetics, Federal University of Rio Grande do Sul,
Porto Alegre, Brazil

Abstract

Bioinformatics is a new field of research, with roots in computer science, physics, statistics and molecular biology. Its growth was fueled by the genome sequencing initiatives and the great amount of data produced by them, having quickly spread to all other fields of biological research. This innovative way of making science proved to be especially useful to immunology, allowing the development of a range of tools to predict critical steps in the immune response. The Immunological bioinformatics, or just Immunoinformatics, has emerged as a promising field, with possible applications in cancer research and vaccine development. Most of these immunoinformatics tools and databases are based on sequence analysis, which brings advantages, but also some limitations. Bioinformatics, however, has already developed ways to deal with tridimensional structures, allowing studying protein-protein interaction at an atomic level and extract, from each amino acid in these proteins, much more information than that provided by the linear sequence. In this chapter, we will discuss new uses of structural bioinformatics to study immunological problems, focusing on the Major Histocompatibility Complex (MHC). The MHC molecule plays a key role in the cellular immune response, presenting peptides derived from intracellular proteins to Cytotoxic T Lymphocytes (CTLs). Through the interaction between these peptide:MHC complexes (pMHCs) and the T Cell Receptors (TCRs), these CTLs are capable of identifying non-self peptides and eliminate infected or transformed cells. Structural bioinformatics tools, such as Molecular Docking and Molecular Dynamics, are being used to build the tridimensional structure of unknown pMHC complexes – not yet determined by classical experimental methods – and to assess molecular aspects of these structures. For instance, the electrostatic potential of the TCR-interacting surface of these pMHCs, as well as Accessible Surface Area (ASA) of the peptide inside the MHC cleft, can be related to the immunogenicity of a given virus-derived epitope. Moreover, these features can be included in one global analysis that can cover most of many details underlying TCR/pMHC interaction, affording to predict the impact of viral escape mutations and selection of epitopes with potential to induce cross-reactive immune responses.

Immunology, Vaccinology and Bioinformatics

Immunology and vaccinology are close related sciences, and advances in one field directly affects the other. The very idea of a vaccine – something that you give to an organism, such as part of a pathogen, in order to prevent a disease caused by this agent – is only possible due to the existence of an adaptive immune response and the mechanisms of immunological memory. In the same way, vaccinology and immunology benefit from microbiology and other sciences, since the knowledge of the pathogen biology can help us to create innovative ways to trigger immunological defenses. Science, however, is not always a linear and logical line of steps. The first well succeeded "vaccination" against a virus was performed by Edward Jenner in 1796 [1], near a century before the first description of a virus [2]. The historical distance for a more refined description of the adaptive immunity is even greater: The Major Histocompatibility Complex (MHC), a key molecule in cellular immune response, was just described in the 1940s [3], and the genetic mechanisms behind antibody diversity was only enlightened in the 80s [4, 5].

Even without knowing, Jenner has also performed the first reported case of induced heterologous immunity [6]: He used a virus that circulates in cows (*cowpox virus*) to immunize against other agent, *smallpox virus*, which was responsible for causing an important disease in humans - called Smallpox or Variola. Since it is not well adapted to humans, *cowpox* cannot cause the disease. However, it is so similar to *smallpox* (in an antigenic way) that is enough to stimulate a protective immune response against this human pathogen.

Vaccinology has evolved enormously with the development of microbiology, as well as with the use of current-available molecular biology tools. Paradoxically, despite our unquestionable success in eradicating *smallpox* and preventing several other diseases, we have actually effective vaccines for a just a small number of pathogens[1]. Of note, in many cases, the most effective way to immunize against a pathogen involves an approach very similar with that of Jenner: the use of "live" attenuated agents. This, however, is not a viable option with viruses such as HCV and HIV, since there is always a risk associated with "live" viruses (capable of replicate and recombine). On the other hand, vaccines using "killed" micro-organisms and subunit antigens are much safer, but normally less effective in inducing protective immunity.

During the host organism infection, a virus will trigger the two main branches of adaptive immunity: (1) humoral immune response, with the production of antigen-specific antibodies

[1] The complete list of vaccines licensed for immunization and distribution by the FDA in the USA can be found at http://www.fda.gov/BiologicsBloodVaccines/Vaccines/Approved Products/ UCM093833.

released in the circulation, and (2) cellular immune response, based on Cytotoxic T Lymphocytes (CTLs) capable of identify and eliminate infected cells. Since "live" attenuated agents will infect cells almost in the same way of a wild-type virus, they will also trigger both responses. Subunit antigens and "killed" micro-organisms have good effect over antibody production, which is triggered independently of cell infection, but normally present poor induction of cellular immune response.

Design new vaccines capable of stimulating the production of memory CTLs, without the need of producing attenuated strains, is clearly an important goal in vaccinology. Despite the fact that predict and manipulate CTL responses remains as a challenging task, immunologists have made important advances, with a growing understanding about the molecular and cellular mechanisms driving CTL stimulation. For instance, the main pathway responsible for presenting endogenous-derived peptides to CTL recognition has already been described in details. This "endogenous antigen-presenting pathway" is an important mechanism of control over the cytosolic proteins, being performed by all nucleated cells in the organism. Through this pathway, a sample of the cytosolic proteins is targeted for degradation and part of the derived peptides will be presented at cell surface. During the infection, the virus use cellular machinery to produce viral proteins that will start to accumulate in the cytosol. Inevitably, some of these proteins will also be degraded, and virus-derived peptides will end up being presented at the surface of the infected cell. Therefore, this pathway is essential to stimulate specific populations of CTLs, as well as to indicate infected cells, which will be recognized and eliminated by activated Cytotoxic T Lymphocytes.

Interestingly, this pathway has many systematic steps, which operate under specific sets of rules. For instance, the degradation of the targeted protein will be performed by an enzymatic complex, which has differential probability to cleave in determined sites of the protein. The generated peptides will be then transported to the Endoplasmic Reticulum (ER) through a specific transporter, which has greater affinity for some groups of peptides with specific amino acids. Finally, the remaining peptides will be bound to the molecule responsible for the presentation at the cell surface (the already mentioned MHC). However, this molecule has also several "requirements", which will be filled by just a small part of the peptides. Despite a given virus-derived protein could theoretically generate a great number of small peptides (by different combinations of the points of cleavage), just a small part of these theoretical peptides will actually be presented to the CTLs and, therefore, just this small fraction of peptides will actually be able to trigger the immune response.

Despite all biological variability, if we know the rules driving each one of these steps we can theoretically reproduce these processes through mathematical and computational models. In addition, the sequencing of human and other model organism genomes has produced a large bulk of data, which has been deposited in several databases, many of which specifically intended to compile immunological relevant data [7]. Bioinformatics was already a well-established science, developing innovative ways to store and analyze all sorts of biological data. This new rising branch of bioinformatics - focused on immunological data - was named Immunoinformatics [7-9]. Following the development of dedicated databanks, many web-based methods were developed to explore this stored data – initially constituted mostly of gene and protein sequences. Refined computational methods were also developed to simulate and predict several important immunological mechanisms, such as the steps of the endogenous antigen-presenting pathway. Nowadays, knowing the amino acid sequence of a given virus-derived protein and some information about the host (e.g. MHC alleles), it is possible to predict with high confidence which peptides will be preferentially presented at the cell surface. These tools and databases have direct application to vaccine development, and some success has already been achieved [9-11].

In this chapter, we will focus on the MHC class I and the endogenous antigen-presenting pathway. However, there is also another class of MHC which is presented only by professional antigen presenting cells (APCs), such as macrophages, dendritic cells and B cells. The MHC class II is involved with the exogenous antigen-presenting pathway, being also an important target for immunoinformatics studies.

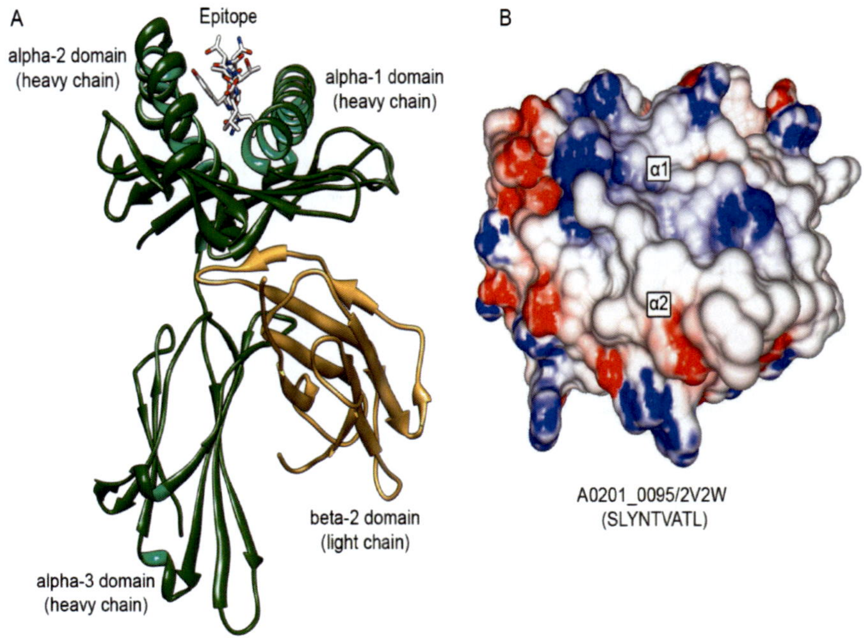

Figure 1. The structure of a pMHC complex. (A) Side view of the tridimensional structure of a MHC-I heterodimer molecule, represented as *cartoon* with the UCSF Chimera package [14]. The heavy chain (α chain) is depicted in green and the light chain (β2-microglobulin) is depicted in mustard. The ligand (epitope) is also represented inside the MHC-I cleft, with colors indicating each atom (*CPK*). (B) Top view of the "TCR-interacting surface" of a pMHC complex presenting an HIV-derived epitope (sequence SLYNTVATL) in the context of a human MHC allele (HLA-A*02:01). The α1 and α2 domains, which delimit the MHC binding cleft, are indicated. CrossTope DB and PDB reference codes are also depicted. Colored areas indicate charge distribution over the surface, in a range from -10 kiloteslas (red) to +10 kiloteslas (blue). Image obtained with the GRASP2 program [15].

Sequence analysis of host and pathogens proteins can answer several important questions and greatly contribute to vaccine development. However, there is still some unpredictable variability that remains and, unfortunately, ends up preventing a larger applicability of this immunoinformatics results. After being processed, the virus-derived peptide will be presented at the cell surface, as part of a complex with the MHC (peptide:MHC or pMHC complex) (Figure 1A). Despite we can predict peptides most likely to be presented in the context of a given MHC allele, the recognition of this pMHC complexes seems to be somewhat more complex than we can predict just looking at the epitope amino acid sequence. Although some clues can be given by the amino acids chemical properties [6, 12], cross-recognition events and heterologous immunity are really difficult to predict. For instance, even single conservative amino acid substitutions in the presented epitope can abort the CTL recognition of a given pMHC [13]. With the recent advances in structural bioinformatics and the fast ever-growing improvements in the computational power, we can now explore the pMHC recognition problem through a structural perspective, which may help us to explain part of the remaining questions and, hopefully, to improve our ability to develop antiviral vaccines.

Structural Databases for Immunological Research

The immune system is constantly engaged in immunological responses, regardless the existence of an established pathological infections [16]. Its mechanisms are needed for a broad range of activities, such as control of our intestinal biota, self-nonself recognition, among many others. For teaching purposes, different immune mechanisms can be roughly divided in some main branches, such as innate and adaptive immunity. While the former involves physical and chemical barriers, and mechanisms specific to more general patterns, such as LPS (lypopolissacaride), the second refers to a more flexible set of responses. Adaptive immunity can be subdivided in humoral response (through the production of antibodies by plasma cells) and cellular immune response (through the CTLs and the MHCs), both of which will be specific to the triggering antigen. During an actual immune response, however, all these mechanisms will act together, and the many interactions between cells and molecules from these main branches highlight the fact that these didactic divisions are not so clear *in vivo*.

One of the biggest challenges of immunologists is to uncover the molecular features underlying one of these specific cell-cell interactions: the recognition of a pMHC by a T Cell Receptor (TCR). As briefly introduced in the previous section, this is a major step in the cellular immune response and, therefore, the focus of many studies related to immunity to cancer, autoimmunity, transplants and vaccine development. There are some important databanks hosting sequence information about MHC alleles and the several peptides presented in their clefts. In this context, stands out the Immune Epitope Database and Analysis Resource (www.immuneepitope.org)[17]. The IEDB contains data related to antibody and T cell epitopes for humans and other species, with information related to many infectious diseases. It also contains experimental data regarding to MHC binding and T Cell assays, as well as several tools to simulate steps of the endogenous antigen-presenting pathway (for a more detailed discussion about IEDB and other related databases, see Rigo et al. 2011 [18]). However, as already discussed, sequence information may not be enough to explain all the variability observed in pMHC/TCR interaction. In order to provide additional information about this issue, new databanks hosting tridimensional structures of pMHC complexes were developed, allowing performing several interesting analyses. In this section, we will focus on three of these structural databanks, showing how to perform the search and which information we can extract from each one of them.

IMGT/3DStructre-DB

This databank is part of IMGT (The international ImMunoGeneTics information system®), a high-quality integrated knowledge resource specialized in the immunoglobulins, TCRs, MHCs of human and other vertebrates, and related proteins of the immune systems that belong to the immunoglobulin superfamily and to the MHC superfamily [19, 20]. The IMGT/3Dstructure-DB hosts tridimensional structures of immune system molecules; protein complexes, such as pMHCs, TCR/pMHCs and others from related proteins from immune system (RPI). All structures come from the Protein Data Bank (PDB) and sequence information is obtained through alignments between the FASTA sequences of 3D structures and the reference sequences from IMGT domains. Genes are then classified according to Human Genome Organization (HUGO) approved nomenclature.

The IMGT/3Dstructure-DB offers several types of searches. It's possible to use an *Entry code* (PDB, INN or protein) or the molecule name. It's also possible to search by a complex, through four different options: (i) *Paratope/epitope*, where one can select IG/Ag, TR/pMH1, TR/pMH2, RPI/pMH1 or RPI/pMH2; (ii) *Peptide/MH*, with the options *pMH1* and *pMH2*; (iii) *Ligand category*, where is possible to chose among *any, chemical compound, DNA, nucleotide, peptide, protein* and *RNA*; (iv) *Peptide length*. There is yet the search by IMGT entry type using IMGT-ontology concepts, the search by resolution, release date or experimental method, the search by bibliographical references and, finally, chain alignment.

Search results are presented in a table containing the following fields: IMGT entry Id, IMGT molecule name, IMGT entry type, IMGT receptor description, species, ligand(s), experimental technique, resolution, PDB release date. By left-clicking at the IMGT Id of the complex of interest it's possible to access several detailed information, including chain length, chain description, chain sequence. It's also possible to access the contact analysis, in which are presented information about pMHC contact sites, describing the MHC residues that interact with the peptide side chains and the interaction type (polar, hydrogen, nonpolar). Besides peptide:MHC interactions, there is also information regarding to TCR/pMHC interactions.

Another interesting approach is use the 3D view. It allows to visualize the protein structure and, by left-clicking in the *IMGT pMH contacts sites* option, to identify the contacts between a given MHC and a given peptide, for instance.

MPID-T2 Database

The MPID-T2 database hosts curated data with information related to 3D structures of pMHC and TCR/pMHC complexes. The database currently hosts 415 entries of MHC molecules from different sources: 282 human, 127 murine, 3 rat, 2 chicken and 1 monkey. Search can be performed by categories, with the following options: *MHC class, organism, dataset, MHC allele, peptide length, output* and *TR type*. There is also the search by *PDB data*, in which one can use the PDB ID, resolution or release year. Each structure deposited in the databank is manually curated and analyzed under several aspects. All the information provided by each one of the complexes hosted at the MPID-T2 database is listed in Box 1.

As can be seen in Box 1, this database extracts much information about the complexes, such as hydrogen bonds (calculated through the program HBPLUS [21]), interface area (through NACCESS program [22]), gap volume between the peptide and the MHC binding cleft residues upon pMHC complex (calculated using SURFNET [23]). Other relevant information such as image plots of complexes interactions generated by LIGPLOT program [24] and image of electrostatic potential of pMHC and TCR interfaces are also available. It's also possible to access the graphical representation of the alignments using the Weblogo software [25].

Box 1. Information provided by MPID-T2 database for each hosted pMHC and TCR/pMHC

MPID-T2 ID	MHC Type	MHC Allele
PDB ID	MHC Source	Peptide Source
TR Source	Experimental Method	Peptide Length
pMHC Structural Alignment	pMHC Interface Electrostatic Potencial	Peptide Sequence
pMHC Interface Area	pMHC Gap Volume	pMHC Gap Index
pMHC Binding Energy	pMHC Contact Area	TR/pMHC Interface Area
TR/pMHC Gap Volume	TR/pMHC Gap Index	TR/pMHC Binding Energy
TR Docking Angle	TR/pMHC Contact Area	TR Type
TR/pMHC Structural Alignment	Resolution	Release Year
Total Hydrogen Bonds in pMHC	Total Hydrogen Bonds in TR/pMHC	TR α-Chain Hydrogen Bounds
TR β-Chain Hydrogen Bonds	MHC Chain ID	Peptide Chain ID
TR α-Chain ID	TR β-Chain ID	pMHC Complex
TR/pMHC Complex	TR Interface Electrostatic Potential	pMHC Ligplot
TR/pMHC Ligplot	Links	Reference

CrossTope Data Bank

Cytotoxic T Lymphocytes (also known as CD8+ cells) circulate in our organism exploring the surface of nucleated cells, through the interaction of its T Cell Receptor (TCR) with pMHC complexes presented in the cells surface. T Lymphocytes pass through a hard selection process in the thymus, so that only CTLs with TCRs capable of identifying non-self peptides presented by self MHC alleles (in "the context" of self MHCs), are released into the bloodstream. Therefore, if a given cell is infected by virus (or other intracellular parasite), the virus-derived peptides will be presented at the cell surface in the MHC context. These pMHC complexes presenting non-self peptides will be recognized through an specific TCR, and the infected cell will be eliminated by the cytotoxic activity of the CTL. Of note, a given CTL will explore a great number of pMHCs (presenting both self and non-self peptides), but will specifically recognize just a small part of these pMHCs. Considering the possible application to vaccine development, it is of great interest to uncover the molecular features responsible for cross-reactivity, that is, which shared characteristics allow two unrelated pMHCs to be recognized by the same TCR. As previously discussed, sequence analysis alone cannot explain or predict all the variation observed in TCR/pMHC interaction. Structural analysis could provide additional insights into this interaction, but there are yet a small number of pMHC complexes available at the PDB. In this context, the CrossTope Data Bank arises as the first repository for *in silico* predicted pMHC complexes presenting immunogenic epitopes. The 3D structure of these complexes was obtained through a combined approach of Docking and Energy Minimization, previously described as *D1-EM-D2* (*Docking 1 - Energy Minimization - Docking 2*)[26]. This approach will be also discussed below.

All immunogenic peptides included in the databank were manually curated from the literature, by experts in the field. Currently, the databank hosts 96 modeled complexes with human MHC alleles (HLA-A*02:01 e HLA-B*27:05) and 85 modeled complexes with murine alleles (H2-K^b e H2-D^b). Besides, CrossTope also hosts a number of crystal structures of immunogenic pMHC complexes derived from the Protein Data Bank.

The user can perform a search by MHC or by peptide and the output will be a list containing information regarding to sequence of the epitope, complex code, source protein, source organism and a "compare" tool, at the right (Figure 2A). By left-clicking at the "plus button" it is possible to obtain more detailed information about each complex (Figure 2B). In addition to several information about the complex and the epitope, each CrossTope entry provides crosslinks for other databanks, such as NCBI, IEDB and PUBMED.

Figure 2. The CrossTope DB output. (A) List of identified complexes in a CrossTope search. (B) Information available for each pMHC complex hosted in CrossTope Data Bank.

The pMHC structure can be visualized and analyzed directly in the website, using Jmol. The user can also download the PDB file of the complex to perform other analysis. Images of the electrostatic potential of the TCR-interacting surface of each complex (generated through the GRASP2 program) can also be downloaded. Of note, the "compare" option (Figure 2A) allows the selection of two complexes of interest to compare the images of electrostatic potential, as well as the other information fields of the complexes. Figure 3 presents the comparison of two cross-reactive TCR-interacting surfaces, from pMHCs presenting an influenza-derived peptide and an herpesvirus 4 derived peptide. Although the sequences of the peptides present only two conserved residues, the electrostatic potential of both complexes is quite identical, which could justify the cross-recognition of these two complexes by the same T Cell population.

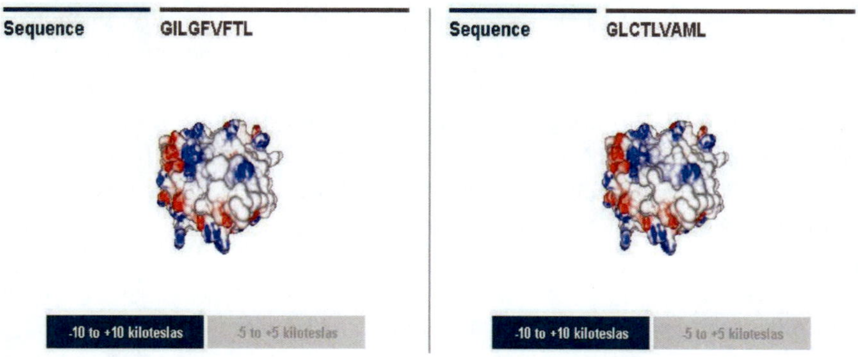

Figure 3. Using the "Compare" tool of Crosstope DB. As an example, the comparison between electrostatic potential images from two cross-reactive peptides presented by HLA-A*02:01, is shown. On the left a peptide from Infuenza A virus and, on the right, a human herpesvirus 4 derived epitope.

The Need of Alternative Methods for Three Dimensional Structure Predictions

The classical experimental methods used to determine the tridimensional structure of biological macromolecules, mainly proteins, are Nuclear Magnetic Resonance (NMR) and X-Ray Crystallography. Many recent advances in these methodologies have allowed an amazing growing of the number of determined structures deposited in the major structural databanks. Protein Data Bank (PDB), for instance, has almost doubled the number of structures in the past five years, going from 40,624 entries in 2006[2] to more than 76,000 structures nowadays[3]. Although the expectations indicate that these numbers will keep increasing in the near future, this rate is far from being enough to keep up with biological variability. First, these methodologies are extremely expensive and time consuming, making them unpractical choices to determine the structure of close related proteins, such as mutated variants and genetic alleles. The most general contribution of the many works being conducted to

[2] Data from RCSB PDB statistics. Available at http://www.rcsb.org/pdb/statistics/contentGrowthChart.do?content=totalandseqid=100.

[3] As of Tuesday Sep 20, 2011 at 5 PM PDT there were 76003 Structures at PDB. For an updated number, visit http://www.rcsb.org/pdb/home/home.do.

determine protein structures through classical experimental methods has been to perform a large screening over all kinds of protein folds and motifs, providing reliable templates for whatever sequence that one wants to model. Second, biological variability is outstanding, and there is an unlimited universe of structures waiting to be determined and studied.

The Major Histocompatibility Complex region in the short arm of human chromosome 6, for instance, is the most variable region of the human genome. Among other immunologically important proteins, this region encodes the heavy chain of three different complexes responsible for the presentation of cytosolic-derived peptides to CTLs (the final step of the already presented "endogenous antigen-presenting pathway"). These three Human Leukocyte Antigen genes (HLA-A, HLA-B and HLA-C)[4] have already more than 5,000 alleles described in human populations[5], of which very few have already known structures. As already discussed in previous sections, these proteins are extremely important to the maintenance of cellular immune response and, therefore, to antiviral immunity. The fully functional structure of an HLA is in fact an heterotrimer, composed by a heavy chain (encoded by one of the HLA genes), a light chain (the extremely conserved β2-microglobulin) and one small peptide (with 8 to 11 amino acids of length) lodge in a cleft between two domains of the heavy chain. This peptide:HLA complexes (or pMHCs) will be transported to the cell surface, in order to interact with the T Cell Receptors (TCRs). Through the TCR, the lymphocyte is able to interact with this merged surface of peptide and HLA (Figure 1B, 2B and 3), identifying non-self peptides presented by the HLA alleles of a given individual. In case of viral infection, virus-derived peptides will be presented in the surface of the infected cell, allowing the immune system to eliminate this cell and to mount an immunological memory against this viral serotype. However, variants of this agent can carry mutations that exchange certain amino acids over the presented peptide, what might be enough to abort the recognition by the memory T cell. Therefore, in order to study cellular immune response against viral scape mutations and to develop effective vaccines against different serotypes, affecting different human populations, immunologists might want to know not only the structure of those 5,000 HLA alleles, but also the structure of millions of different complexes presenting the thousands of peptides that can be presented by each one of these alleles. This example makes it clear that we might not ever have enough resources to determine all biologically interesting structures through classical experimental methods.

The expression "MHC alleles" is used to refer both the genes and the proteins encoded by these gene variants. However, this expression is not quite correct once "allele" refers to DNA, and not to protein. The expression allotypes is indicated to refer to proteins defined by different MHC alleles [27], and will be used in the following discussion (the expression "MHC alleles" will still be used to refer to MHC gene variants). Nowadays, there are less than 400 structures of pMHC complexes available[6] (considering all organisms), many of which are redundant (same allotype and the same epitope). A search at IMGT/3Dstructure-DB for the HLA-A*02:01 allele, which has the greatest number of structures available, returns only 152 PDB codes. But what are the viable alternatives? Using the pMHC example,

[4] The heavy chain of the complex responsible for presenting peptides at the cell surface is called Major Histocompatibility Complex (MHC), but this name changes according to the organism. For instance, in humans it is called Human Leukocyte Antigen (HLA), in mice is called H2 antigen, etc.

[5] As of Sunday Nov 20, 2011 there were 5,468 MHC class I alleles listed at http://hla.alleles.org/ nomenclature/ stats.html.

[6] As of Tuesday Sep 27, 2011 there were 396 MHC class I structures available at IMGT/ 3Dstructure-DB.

what are the available tools to determine and study the tridimensional structures of these innumerous complexes? The answer for these questions relies on bioinformatics tools, and we will here divide the focus in two smaller problems: (1) Determining the unknown structure of an HLA allotype and (2) constructing pMHC complexes through *in silico* approaches.

Revealing the Structure of New HLA Allotypes through Molecular Homology Modeling

MHC class I molecules (MHC-I) are responsible for the presentation of endogenous-derived peptides at the cells surface. To perform this task, MHC class I molecules have a much conserved structure. All MHC-I are heterodimers, with one light chain (β2-microglobulin) and one heavy chain (α), which presents three different domains (α1, α2 and α3). Domains α1 and α2 interact to form a cleft capable to accommodate peptides with 8-11 amino acids long. The heavy chain is encoded by three different genes (HLA-A, HLA-B and HLA-C) located at the most polymorphic region of the human genome, the MHC Region at the short arm of chromosome 6 (6p21.3) [28, 29]. Together, these three MHC class I human genes have already 5,468 described alleles[5]. The light chain, on the other hand, presents a conserved chain encoded by a different locus, at the long arm of chromosome 15 (15q21-q22.2).

The general structure of the MHC-I molecule is quite conserved. Despite all variability observed at the heavy chain, which will have consequences over peptide specificity and TCR recognition, all MHC-I molecules will have a cleft formed by a platform of β-sheets, which supports two parallel α-helixes (even in other species). As already discussed, there are very few MHC allotypes with available 3D structure at PDB, and even few amino acid changes can significantly alter the environment of the binding cleft. The conserved general structure of the heavy chain, and the great number of available genotyped sequences, makes the MHC-I heavy chain an important target for molecular homology modeling.

Different approaches were already applied to build MHC-I homology models. Cardozo at al. (2005) proposed the use of the ICM method with a Biased Probability Monte Carlo (BPMC) procedure [30]. As briefly summarized by Bordner et al [31], the method consists of the following steps: (1) Calculate a global alignment, (2) build an extended polypeptide chain, (3) assign torsion angles for the aligned portions of the backbone, using the same torsions for the side chain of residues identical to the template structure, (4) assign the most likely rotamer to the remaining residues (that has no template for the side chains). Finally, (5) they iteratively minimize a sum of the physical energy and quadratic restraints between corresponding atoms in the model and template structures. This approach was used to build MHC structures from both human and murine allotypes.

In 2010, Khan et al [32] proposed an automated procedure for pMHC structure prediction (pDOCK) which includes an step of homology modeling of the receptor by satisfaction of spatial restraints for MHC sequences (pDOCK will be discussed in the next section). The authors also discuss that high quality MHC-I homology models can be generated using a previously described three-step homology modeling procedure, which uses the MODELLER program [33].

One of the most important questions about homology modeling is how closely related are your template and your target. Homology modeling tools will try to "fit" your target sequence into your template structure, and after refine the target structure according to steric and chemical restraints.

But as already discussed, even few amino acid substitutions can affect the cleft topology. Could homology modeling produce reliable MHC structures even using as template an MHC allotype with a very different specificity? Recently, Cárdenas et al. [34] have used the Swiss-Model tool of the ExPAsy server to build 3D models of pMHCs with Salmon (*Salmo salar*) MHC allotypes, using as template the structure of a murine allotype (H-2K^b).

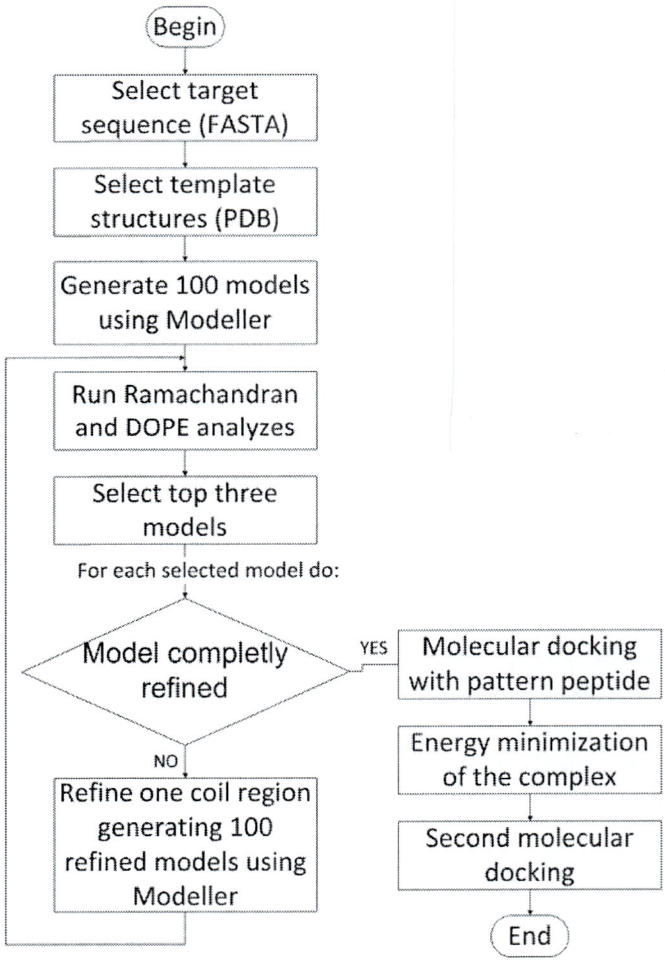

Figure 4. Dataflow diagram of modeling process and evaluation. A 100 models are generated with Modeller software using related PDB structures as templates. A combined analysis of Ramachandran plot and DOPE score are used to select the best models (PROCHECK program). Coil refinement rounds are performed with the best models, as necessary. After all stereochemical parameters are satisfied, a "functional test" is performed by docking an epitope (restricted to the MHC allotype that is being modeled) into the modeled cleft. This final test was performed using the *D1-EM-D2* approach (as discussed below).

In this study, the modeled complexes were submitted to further evaluations and to a molecular dynamics simulation. Even though the modeled structure presents good values for all structural and stereochemical descriptors, it is quite difficult to assure that a given model is not too much tied to the template structure.

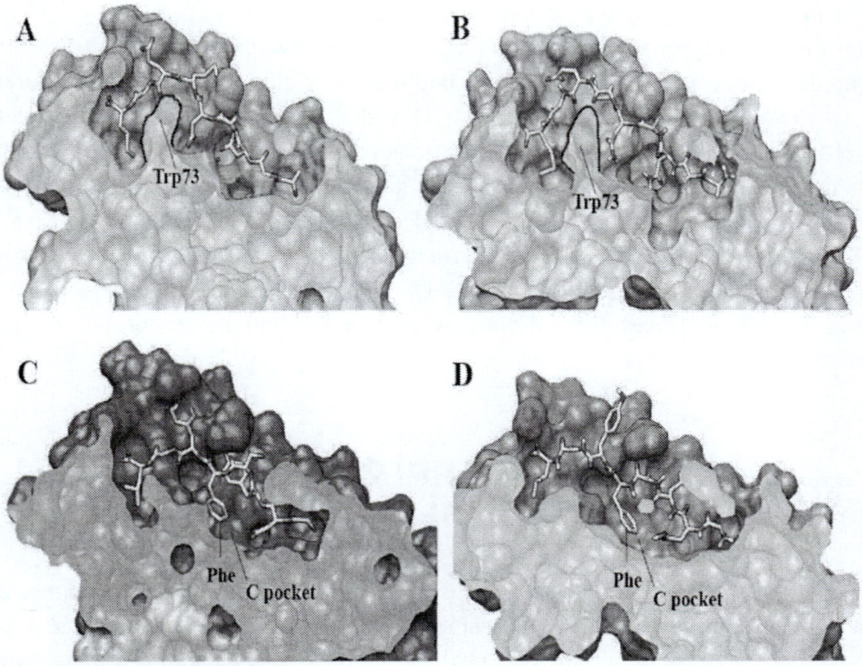

Modified from Fülber et al. 2010 [38].

Figure 5. Side view of pMHC clefts. (A) 9-mer:H2-D^b crystal (1HOC), (B) 9-mer:H2-D^b model, (C) 8-mer:H2-K^b crystal (1LK2), (D) 8-mer:H2-K^b model. The bulge at Trp73, in both the H2-D^b model and crystal, are depicted. The deeper C pocket of H2-K^b is also indicated.

The two most studied MHC murine allotypes are H2-D^b and H2-K^b. H2-D^b molecules present specially 9-mers and 10-mers, and its anchor positions are p5 (normally an asparagine, "N") and p9 (usually M, I or L). On the other hand, H2-K^b usually presents 8-mers, and its anchor positions are p5 (F or Y) and p8 (L, M, I or V) (anchor residues for several allotypes can be verified at http://www.syfpeithi.de [35]). Differences between peptides restricted to these allotypes can be partially explained based on the pattern of hydrogen bounds generated between the peptide and the MHC molecule in each allotype [36] and specially by the presence of a tryptophan residue (Trp73), in the D^b molecule [37]. This residue from α1 domain is facing another tryptophan from the α2 domain (Trp147, which is conserved in several MHC allotypes) and together they almost close the H2-D^b cleft, forcing the peptide to pass above them. H2-K^b molecules have a serine at this site (Ser73 of α chain), which results in a deeper cleft in this region (pictures of both clefts and a more detailed structural comparison can be found in Antunes et al. (2010) [26]).

Aiming to evaluate the ability of an homology modeling protocol with MODELLER program to build a reliable MHC model using a quite different MHC allotype as template, Fülber et al (2010) performed a cross modeling of these two murine allotypes [38]. H2-K^b

alpha-chain templates were used to build a H2-D^b model and, conversely, an H2-K^b model was obtained using H2-D^b alpha-chain templates. More than a hundred models were generated for each allotype, followed by loop refinement rounds, as necessary (Figure 4). Evaluation of stereochemical and structural quality for all models was carried in Procheck v.3.5 and results of Ramachandran plot and DOPE score were evaluated together. But since these two murine allotypes have different specificities, it was important to find a way to verify the functionality of the modeled clefts. In order to evaluate these functional aspects, a molecular docking was performed, where peptides restricted to each MHC allotype were docked to the respective modeled MHC (GNYSFYAL for H2-K^b; ASNENMETM and SQLKNNAKEI for H2-D^b). Docking calculations presented good values of Binding Energy for both models and a qualitative analysis of the MHC cleft highlighted the similarities between the models and the reference crystal structures. As expected, the H2-K^b model presented a deeper pocket, admitting the arrangement of the anchor residue, in this case a phenylalanine in the fifth epitope position. On the other hand, H2-D^b cleft presents the tryptophan 73, which forces the epitopes to assume the pattern conformation for ligands presented by this type of MHC (Figure 5).

In Silico Approaches to Construct pMHC Complexes

Even for allotypes with already known tridimensional structure, *in silico* construction of different pMHC complexes – with the same allotype presenting several other peptides of interest – remains as an important and challenging field of research. Despite some MHC allotypes have already few tens of different structures available, such as the human allotype HLA-A*02:01 and the murine allotype H2-D^b, there are countless different peptides that we want to analyze in the context of these allotypes: virus-derived peptides from many serotypes, cross-reactive epitopes, self and non-self peptides involved in autoimmunity, cancer-derived peptides, epitopes from intracellular bacteria, and the list goes on.

Several different strategies were used to address this issue, as well reviewed by Bordner and Abagyan, 2006 [31]. Many of these approaches are based on molecular docking, a well-known structural bioinformatics tool largely used to predict the conformation of small ligands inside the binding site of a macromolecule (a receptor or an enzyme). But while most ligands used in virtual screenings have up to 10 rotatable bonds, a typical MHC-I epitope has about 9 amino acids, possible with some long side chains and more than 40 rotatable bounds to account. Predict the best fit of such a complex structure inside an MHC cleft is not an easy computational task. In order to avoid dealing with all these variables at once, most strategies involve some kind of assumption, or even splitting this problem in two or more smaller tasks.

MHCSim Server

The first web server to automated construction of pMHC complexes - MHCSim - became available in 2007. It was part of the ImmunologyGrid, a project to use GRID computing for *in silico* prediction of binding affinities between MHC molecules and associated peptides. The MHCSim service was able to build up pMHC complexes with several human MHC allotypes (HLA class I and HLA class II), providing the desired structure almost instantaneously. However, the generated file was not supposed to hold the final trustworthy structure of the desired peptide:HLA. Instead, it is actually an initial draft for subsequent refinements and dynamic simulations. The server uses the input sequences (chosen HLA allotype and given peptide sequence) to search in an internal structural database (supplied with PDB structures) and find the peptide:HLA structure most closely related to the input. This template structure is then mutated in the discordant positions, in order to create the desired structure.

However, simply mutate one structure into another is not enough to achieve the actual peptide conformation. Even one single amino acid substitution can induce some adjustments in the backbone of the peptide, with possible consequences over the side chains of other non-mutaded positions. This adjustment will not be accounted in the structure generated by the MHCSim, but may occur during a molecular dynamic simulation using the provided pMHC coordinates. Molecular Dynamics (MD) is a powerful *in silico* tool to assess dynamic information about the structure, such as stability, solvation degree, secondary structure integrity, among many others (more detailed discussion will be carried later). It is not, however, a tool designed to provide a final structure coordinates file. The final conformation (after the MD) is just a picture in a non-ended movie, and one that intends to obtain only one representative structure of the simulation, will need to extract some reliable approximation, such as the most frequent conformation.

D1-EM-D2 Approach

Analyzing the pMHC crystal structures already available at PDB, our group of research started to identify certain conformational patterns. For instance, at the time (2007-2008), there were 28 available structures of non-redundant pMHC complexes with the murine allotype H2-D^b (Table 1). Fitting these structures, we observed a striking identity of the peptide backbone conformation for all 28 different peptides. Despite all the sequence variability and the different origins of these peptides (some were virus-derived, others were host-derived), all of them adopted nearly the same backbone pattern inside the H2-D^b cleft. For the side chains, conformation varies according the position on the peptide sequence (Figure 6).

Side chains of the anchor positions (amino acids of the peptide primary responsible for holding the peptide inside the MHC cleft) are buried in the same specific MHC pockets and, therefore, also presented almost the same conformation (Figure 6). Side chains of the central positions (mainly p6 and p7), on the other hand, presented completely variable conformations, being almost unique for each peptide. This makes much sense: (1) anchor positions must interact with the MHC (in this case, the same for all peptides), and peptides that don't have amino acids capable of interact with these specific pockets will not be presented by this specific MHC allotype. Central positions, on the other hand, (2) will be exposed out of the MHC cleft, and the diversity among them will determine the specificity in the TCR recognition of each pMHC complex.

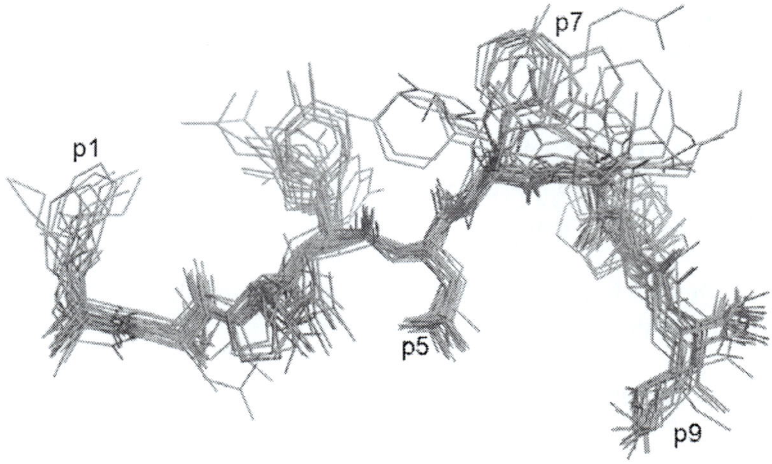

Modified from Antunes et al (2010) [26].

Figure 6. Structural diversity of H-2D^b restricted peptides. Thirty-three files containing H2-D^b allotypes were found in PDB at the time of our search (2007-2008), from which 28 (*) presented different epitopes. These 28 epitopes were superposed for structural analyses. All epitopes share the backbone conformation and there is great conservancy at the buried anchor sites (p5 and p9). On the other hand, great diversity is observed at p7 side chains. 1FFO presents a pMHC interacting with the TCR. Since this interaction could affect the conformation of the epitope, this structure was excluded. Structures presenting incomplete epitope sequence have been also excluded (e.g. 2CII).

In order to verify if peptides presented by different allotypes also shared this (or any) structural pattern, we also analyzed all available structures for other murine allotype (H2-K^b). Most peptides presented by H2-K^b also share a specific conformational pattern for the peptide backbone. However, this pattern was different from the pattern presented by H2-D^b-restricted epitopes (Figure 7). Most important, there were two crystal structures of the same LCMV-derived epitope (KAVYNLATM), one in the context of H2-D^b, other in the context of H2-K^b. Corroborating our observations, the conformation of this epitope in each MHC cleft was in agreement with the described pattern for each allotype.

Differences about the conformation of a given peptide when presented by these two allotypes had been already discussed, but the analysis in a larger scale (with 28 structures of H2-D^b and 21 of H2-K^b) allowed us to unveil general allotype specific patterns. In conclusion, this means that the conformation of the epitope's main chain is not only a consequence of its amino acid sequence, but actually determined by the shape of the MHC cleft, which is specific for each allotype.

As already discussed, prediction of the complete epitope conformation is a difficult computational task, normally splitted in small problems. However, if we know that epitopes presented by a specific allotype will normally present the very same backbone conformation, and we already identified this conformation, as in the case of H2-D^b and H2-K^b, then we can "jump" the main chain prediction step (just assume the pattern) and focus on the resolution of the side chains conformation.

This was the rational for the development of a new approach to build pMHC complexes. Knowing the MHC structure of the desired allotype, and the allotype-specific conformational pattern, we can use any epitope amino acid sequence to predict the tridimensional structure of this specific pMHC complex.

Table 1. List of PDB structures used to identify an H-2D^b^-specific peptide pattern

PDB code	Epitope description	Sequence (aa)	N° of aa
1FG2*	LCMV Peptidic Epitope (gp33)	KAVYNFATC	9
1JPF*	LCMV Peptidic Epitope (gp276)	SGVENPGGYCL	11
1JPG*	LCMV Peptidic Epitope (np396)	FQPQNGQFI	9
1INQ*	MHC Class I H13[a] minor histocompatibility peptide	SSVVGVWYL	9
1JUF*	Minor Histocompatibility Antigen peptide (H13[b]) P4(Val/Ile)	SSVIGVWYL	9
1N3N*	Mycobacterial hsp60 epitope	SALQNAASIA	10
1QLF*	Sendai $NP_{324-332}$ (*gly327 replaced by o-glcnac serine*)	FAPSNYPAL	9
1S7U/2F74/ 1FFN/1N5A*	LCMV-derived immunodominant peptide (gp33)	KAVYNFATM	9
1S7V*	LCMV-derived (gp33) index peptide - escape variants	KAVYNLATM	9
1S7W*	LCMV-derived (gp33) index peptide - escape variants	KALYNFATM	9
1S7X*	LCMV-derived (gp33) index peptide - escape variants	KAVFNFATM	9
1FFO*	Gp33 syntethic peptide with substitution K1A /C9M	AAVYNFATM	9
1FFP*	Gp33 syntethic peptide with substitution K1S /C9M	SAVYNFATM	9
1WBX*	Peptide from Influenza A (pr8) $HA_{468-477}$	SQLKNNAKEI	10
1WBY/1YN6*	Peptide from Influenza A (pr8) $PA_{224-233}$	SSLENFRAYV	10
1YN7*	Mutated peptide (R7A) from RNA polymerase subunit P2	SSLENFAAYV	10
1ZHB*	9-mer peptide from Dopamine beta-monooxygenase	KALYNYAPI	9
1BZ9*	Synthetic Peptide (P1027)	FAPGVFPYM	9
1CE6*	Sendai Virus Nucleoprotein Peptide ($NP_{324-332}$)	FAPGNYPAL	9
2VE6*	Altered peptide of Sendai virus (*Photocleavable peptide*)	FAPGNYPAL	9
2CII	Sendai virus nucleoprotein epitope	FAPGNYPAL	9
2ZOK*	9-meric peptide from Spike glycoprotein	ASLWNGPHL	9
2ZOL*	9-meric peptide from Spike glycoprotein - Mutation: W4S	ASLSNGPHL	9
1HOC*	Influenza virus peptide $NP_{366-374}$	ASNENMETM	9
3CPL*	Influenza virus peptide $NP_{366-374}$ with substitution M6A	ASNENAETM	9
3BUY*	Epitope of PB1-F2	LSLRNPILV	9
3CC5*	Nonameric peptide from Melanocyte protein Pmel 17	KVPRNQDWL	9
3CCH*	nonameric peptide murine gp100	EGSRNQDWL	9
3CH1*	nonameric peptide chimeric gp100	EGPRNQDWL	9

Thirty-three files containing H-2D^b^ alleles were found in PDB at the time of our analysis (2007-2008), but just 28 (*) of them presented different epitopes. These 28 epitopes were used to structure and sequence analyses. 1FFO presents a pMHC interacting with the TCR. Since this interaction could affect the conformation of the epitope, this structure was excluded. Structures presenting incomplete epitope sequences have been also excluded (e.g. 2CII). Modified from Antunes et al. (2010) [26].

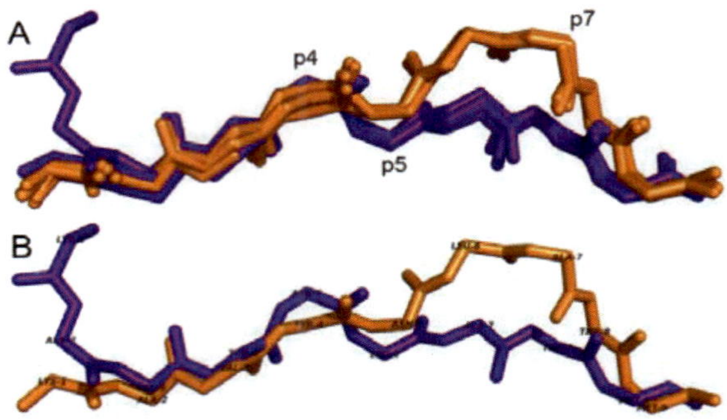

Modified from Antunes et al. (2010) [26].

Figure 7. Superposition of H-2D^b/H-2K^b-restricted epitopes. Structures of five H2-D^b-restricted epitopes (1CE6, 1S7V, 1WBZ, 1ZHB and 3BUY) and of five H2-K^b-restricted epitopes (1FO0, 1FZJ, 1LK2, 1RJY and 1S7R) were superposed. The color pattern was kept on figures A and B. (A) The side view shows the conformational differences among the H2-D^b (orange) and the H2-K^b-restricted epitopes (purple), especially between positions 5 and 7 (p5-7). (B) Two crystal structures (1S7V and 1S7R) representing the same epitope (KAVYNLATM) in the context of these two different allotypes.

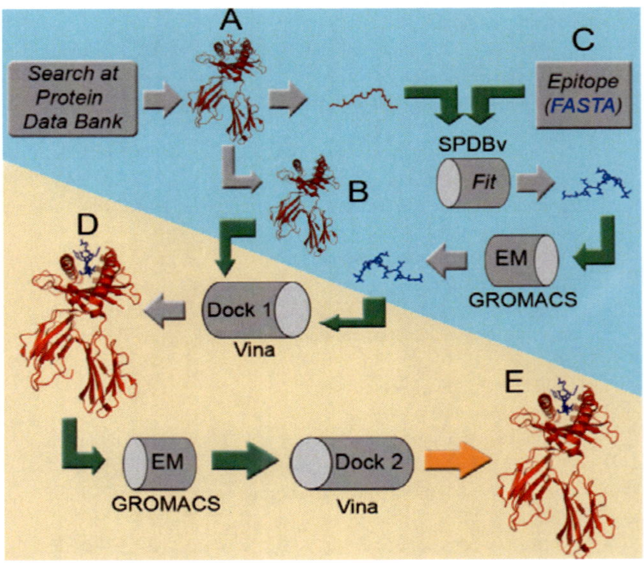

Modified from Antunes et al. (2010) [26].

Figure 8. Dataflow diagram of the *D1-EM-D2* approach. The blue area indicates de preparation steps, and the orange area indicates the *D1-EM-D2* steps. Briefly, a crystal structure of an MHC is used as "MHC donor" (macromolecule) for cross-docking with the structure of a given epitope. Epitope structure is built from its linear amino acid sequence using the crystal structure of another peptide restricted to the same allotype. A molecular docking is performed with AutoDock Vina 1.0.2 [40] using default values. All torsions in the macromolecule are kept rigid, as torsions in ligand backbone, while ligand side chains are kept completely flexible. For each epitope, the same input files were used to run AutoDock Vina 20 times, generating a final population with up to thousand different conformations (20 outputs with up to 50 structures each). Binding energy (BE) values of the best conformation of each output are used to establish a threshold, and the conformations above this threshold are used to calculate the root mean square deviation (RMSD) among the selected conformations. The obtained complex is then submitted to an energy minimization (EM) step, aiming to adjust the "MHC donor" to its new ligand, and the resulting complex is used into a new round of molecular docking. EMs are performed with GROMACS package [41].

For the MHC coordinates, we select one of the PDB structures (with the best possible X-ray resolution) presenting the desired MHC allotype and a given peptide. The peptide sequence of this selected pMHC is not important, since we are only interested in the MHC coordinates. The MHC structure is then saved in a separated pdb file (with an empty cleft), and will be referred as "MHC donor". The desired epitope sequence is used to build an initial epitope tridimensional structure, assuming the allotype-specific pattern for the backbone. In order to predict the best conformation of the epitope's side chains, and to adjust the "MHC donor" cleft to this new epitope, we use three sequential steps: Docking, Energy Minimization of the whole pMHC structure, and a second round of Docking (for that reason called *D1-EM-D2* approach)(Figure 8). During both rounds of molecular docking (performed with Autodock Vina), the MHC donor structure is kept completely rigid, while the epitope's side chains are kept flexible. During EM step with explicit solvent (performed with *Steepest Descent* algorithm of Gromacs package), all the pMHC structure is kept flexible. The goal of the final docking step (D2) is to perform a new search for the best conformation of epitope within this optimized MHC cleft (for complementary information read Antunes et al. 2010 [26]).

Allotype-specific patterns were identified for other MHC alleles (initially the human alleles HLA-A*02:01 and HLA-B*27:05) and this approach was tested reconstructing 46 complexes with already known tridimensional structure. Complexes were successful reproduced with a mean RMSD of 1.754 (± 0.4675) Å for all epitope atoms (including side chains). Reliability of these modeled complexes was also confirmed by visual inspection of the electrostatic potential distribution over the TCR-interacting surface, and through correlation with *in vitro* data of cross-reactivity among specific sets of complexes [26, 39]. This approach was also used to build a large set of so far unknown pMHC structures, which are now available at the CrossTope Structural Data Bank for Cross-Reactivity Assessment (previously presented in this chapter).

The *D1-EM-D2* approach is an easy way to obtain reliable models of unknown pMHC complexes. However, it also has some limitations. The main one is the need for an already known allotype specific pattern and, therefore, for at least some crystal structures of the desired allotype. Pattern for other allotypes are being currently identified, but the coverage of this approach will depend on the continuous input of new MHC structures. Alternatively, it can be used after the determination of the backbone conformation through other *in silico* approaches. The second limitation is related to the universality of the allotype-specific pattern. Although there seems to be almost no exceptions to murine allotype patterns, there are considerable more exceptions for human allotypes, especially regarding to cancer-related peptides [26]. It is important to know that *D1-EM-D2*-derived structures are models, and that some of them can disagree a little with the real structure. Nevertheless, considering all the validation already performed, this approach seems to offer a quite good approximation of the desired pMHC structures.

Ab Initio Prediction of Binding Geometry

Since molecular homology modeling can provide us quite reliable models for new MHC allotypes, to predict the conformation of a given epitope in this cleft remains as a challenging goal in this field of research. Molecular docking is not able to handle alone with all the epitope variability, since the docking accuracy fall down considerably with the increase of the number of rotatable bonds. Alternatively, it is possible to combine steps of molecular docking with steps of structural optimization with Monte Carlo simulation, or even molecular modeling tools. In 2006, Bordner et al. [31] described an approach for *ab initio* prediction of pMHC binding geometry, using molecular docking and Monte Carlo refinement. This method was able to reproduce the epitope's backbone conformation of 23 pMHC crystal structures (14 with HLA-A*02:01 and 9 with H2-K^b), with RMSDs of 1.09 Å for the human allotype complexes and 0.66 Å for the murine allotype complexes. The side chain conformation of anchor positions were also well reproduced, with RMSDs values below to 2 Å. However, this work was primarily focused on predicting the binding affinities through a structural approach, rather than provide accurate models for all epitope side chains (the authors have discussed about this two different goals in pMHC a*b initio* prediction). Since the central amino acids, which will be crucial for the TCR recognition, do not collaborate to the binding affinity, they were not greatly discussed in this work.

This approach was also used to dock peptides inside the cleft of five HLA-B modeled MHC structures, that were obtained by homology modeling using HLA-A*02:01 as template. The modeled HLA-B allotypes already have a few crystal structures, which were used to verify the modeling accuracy. Interestingly, the authors report good results in reproducing the epitope's backbone conformation in the context of most MHC modeled allotypes, and even good results in reproducing the entire epitope structure, in some specific cases.

pDOCK

In September 2010, was published the most recent methodology for peptide docking into MHC molecules [32]. This approach, called pDOCK, was also based on GRID maps (pre-docking calculations) and local minimization steps with Monte Carlo. It was tested through the reproduction of 149 pMHC complexes (Cα RMSD < 2.5 Å). The authors also highlight that 83% of these reproductions presented RMSD values lower than 1 Å, considering only the peptide α-carbons (which is a measure of the backbone conformation). This approach was largely validated, presenting superior results when compared to few previously published alternatives [32]. In addition, the authors also included an optional step of receptor homology modeling (using MODELLER) in the pDOCK automated protocol, which could be applied to MHC allotypes with no available structures. However, despite the fact that this approach includes steps of adjustment of the interaction between each epitope side chain and the MHC cleft, the authors do not provide any information about the deviation presented by the side chains of the reproduced peptides.

Future Directions

A short review about the available *in silico* approaches to construct pMHC complexes can provide us some clues about the advances and the persisting challenges in this field. It is clear the applicability of molecular docking and Monte Carlo refinement to *ab initio* prediction of pMHC structures. Combined with molecular homology modeling, these methods can provide us useful tools to study complex of MHC allotypes with still unknown tridimensional structures. Considering the structural bioinformatics point of view, these approaches were already largely tested and presented exciting results in reproducing crystal structures. However, for the immunology point of view, the question remains about the real applicability of these approaches to answer immunological problems. As already briefly discussed in this chapter, there are two main goals in *ab initio* pMHC predicition: (1) determination of the complex geometry and (2) determination of the peptide binding affinity for the MHC. The epitope backbone conformation, as well the position of its anchor residues, plays a central role in determining the binding affinity between a given peptide and a given MHC allotype. Epitope side chains of other positions, on the other hand, present minor effect over binding affinity. However, side chains of the central amino acids – normally turned out of the MHC cleft – are critical for the specificity of the TCR/pMHC recognition.

Considering these two important structural immunoinformatics goals, there is much more information correlating structural prediction with *in vitro* data regarding binding affinity than studies exploring structural based prediction of immunogenicity (the final effect of a given pMHC over a specific lymphocyte population). Further studies are needed in order to verify the contribution of available structural approaches to predict aspects of the TCR/pMHC interaction, and perhaps helping to uncover the molecular basis of important immunological phenomena, such as the impact of viral scape mutations, molecular mimicry and cross-reactivity.

In this way, our work with *D1-EM-D2* approach to build pMHC complexes presenting epitopes that were previously tested *in vitro*, the gathering of all this immunological and structural information in a new databank (CrossTope), and the structural analyses of these pMHC complexes through multivariate statistical methods (that will be further discussed in this chapter), have provided an innovative way of exploring immunological problems with the available bioinformatics tools. Besides that, our results indicate that structural measurable aspects of pMHC complexes, such as the Accessible Surface Area (ASA) and the electrostatic potential distribution over the TCR-interacting surface, can actually be used to explain variations in the immunogenic potential of virus-derived epitopes. These findings reinforce the importance of *in silico* pMHC prediction and its applicability to future vaccine development studies.

Extracting Immunological Relevant Information from pMHC Structures

Cellular immune response plays a central role in adaptive immunity, being extremely important for vaccine development. On the other hand, the pMHC recognition by the lymphocyte is far the most critical step for cellular immunity. This is actually the final step of a series of events and molecular issues. For instance, a given peptide must be correctly processed and transported to the ER, in order to have chance to interact with the available MHC allotypes. The peptide must have acceptable anchor residues to be bound to the MHC cleft and this complex must be stable to be transported to the cell surface and stay there time enough for the TCR interaction. Finally, this complex will stimulate a given lymphocyte only if it has a specific pattern of topology and charge distribution over the interacting surface. As discussed in recent publication[7], all these steps can be summarized in three selection levels: Processing, Stability and Reactivity. While sequence based methods can provide reliable predictions about the Processing, structural immunoinformatics can provide data about Stability and Reactivity (immunogenicity).

Binding Properties

Binding properties can be evaluated either through quantitative as qualitative methods. Besides *in vitro* studies, e.g. dissociation constant (Kd) of peptide by the MHC, it is also possible to calculate binding energy, for example, from molecular dynamics simulations. But it is only through bioinformatics approaches that we can visualize certain aspects of interactions at molecular level. A software that made this possible is the LigPlot. This software reads a specific file containing a receptor and a ligand (maximum of 10 residues) in the PDB format and computes hydrogen and hydrophobic interactions, allowing the user to visualize the interactions performed with each atom of the ligand. It is also possible to enter two distinct files simultaneously, e.g. the same ligand complexed with two different receptors, so that you can analyze interaction patterns between these two files.

Accessible Surface Area (ASA)

Accessible Surface Area has been applied to molecular dynamics, protein modeling and for virtual screening of small molecules [42, 43]. In the field of immunology, ASA value has already been correlated with immunogenicity [37], and also used to study TCR/pMHC interaction [44]. In a work published in 2008, Fytili et al. tested the influence of each epitope position to the TCR recognition [13]. They create 9 alanine-exchanged variants of the HCV-derived immunodominant epitope $NS3_{1073\text{-}1081}$ (CVNGVCWTV), and tested the immunogenicity of these variants (in the context of HLA-A*02:01) against a wild-type-specific CTL population. This experiment has shown that even single conservative amino acid substitutions could abolish the TCR recognition.

Using the *D1-EM-D2* approach, we built the tridimensional structure of these 9 variants and the wild-type, all in the context of the HLA-A*02:01 allotype [26]. The ASA value of each peptide inside the cleft was calculated and the plot of these values has presented quite identical signatures for peptides that induced strong response *in vitro* (Figure 9A). Nevertheless, a direct correlation between ASA value and immunogenicity was not seen, since some peptides with poor *in vitro* response presented high values of ASA. However, the peptide with the lowest ASA value at p5 (Seq3) was the only peptide that did not stimulate a detectable response *in vitro*.

[7] Chapter being edited for publication by Nova Publishers. Edited Book Title: Major Histocompatibility Complex: Biology, Functions and Roles in Disease. Chapter title: MHC, Viral Infection and Immunoinformatics. Rigo MM, personal communication.

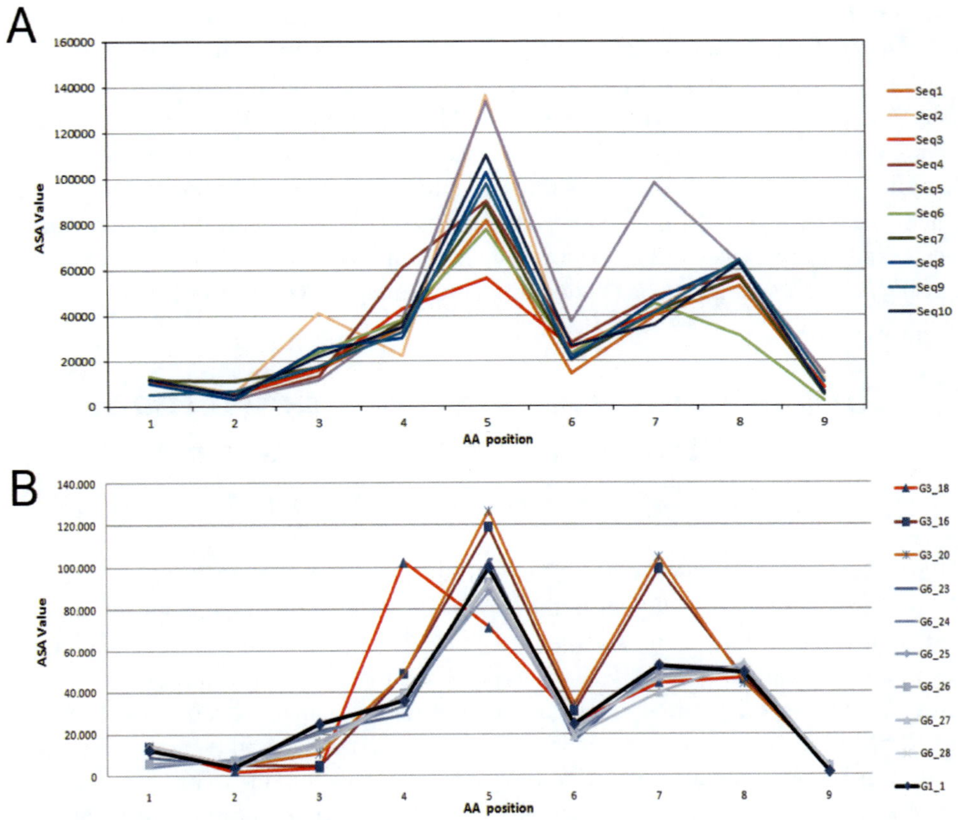

Modified from Antunes et al. 2010 [26].

Figure 9. Plots of Accessible Surface Area (ASA) of different epitopes inside the MHC cleft. (A) ASA values for each epitope residue of the wild type HCV-derived epitope (CVNGVCWTV, S10) and 9 alanine exchanged variants. Variants with high *in vitro* response (S8 and S9) presented almost identical plots, while variants with low response (S1 to S5) presented divergent values, especially at p5. Peptides S6 and S7 presented intermediate *in vitro* response. (B) ASA values of the same wild type HCV-derived peptide (indicated as variant 1 from HCV genotype 1, or G1_1) and 28 naturally occurring variants of this epitope (for the sake of simplicity, values of only 9 variants are depicted in this picture). ASA values from genotype 6 presented the same pattern of the wild type peptide and the peptides with higher deviation from this "ASA pattern" (G3_16, G3_18 and G3_20), presented the lowest levels of IFN-γ production in all ELISPOT assays [13].

In the same work, we also calculate ASA values for another set with 28 HCV naturally-occurring variants of the same HCV-derived epitope, which were also previously tested *in vitro* by Fytili et al. 2008. According to this study, cross-genotype-reactivity was identified between the wild-type epitope (peptide 1 from Genotype 1 or G1-1) and the peptides from HCV genotype 6 (G6-23 to G6-28). In agreement with this data, the same pattern of ASA values per residue was seem for all G6 peptides and the wild-type (Figure 9B). Once again, the peptide with the lowest ASA value at p5 (G3-18) was the only peptide that did not stimulate a detectable response *in vitro*. Moreover, the other peptides with higher deviation from this "ASA pattern" were G3-16 and G3-20, which also presented lowest values of IFN-gamma production in all ELISPOT assays [13].

According to our results, even small differences in ASA values, especially at p5 (central position to TCR interaction), seem to reflect significantly alteration of pMHC recognition. In both datasets, peptides with the highest levels of IFN-gamma production presented plots with a good fit to that of the wild-type. However, we also stress the fact that in the second analysis (bigger dataset) some peptides with low response *in vitro* also presented ASA values similar to the wild-type. Taken together, these results indicate that large deviations in ASA values may indicate a poor stimulation *in vitro*, but the opposite is not true. It is quite clear that ASA values cannot alone explain all immunogenic variation observed *in vitro*, but can certainly be one of the indicators taken into account during a screening for possible targets.

Use of Multivariate Statistical Methods for Structural Virtual Screening of Cross-Reactive Targets

Each Cytotoxic T Lymphocyte (CTL) carries a specific TCR, which can recognize a given *pool* of virus-derived peptides in the context of MHC. The CTL recognizes not only the peptide, but the TCR-interacting surface of the pMHC complex, which is also determined by specific characteristics of each MHC allotype. Therefore, CTLs are referred as being both specific and poly-specific: each CTL can recognize only a small fraction of the universe of possible pMHC complexes (specific recognition), but this specific group of targets can include completely unrelated pMHCs (poly-specific recognition). This mechanism, also referred as cross-reactivity, plays a central role in heterologous immunity and, therefore, in vaccine development.

Cross-reactivity was initially described as being triggered by the great similarity between amino acid sequences of the presented peptides. However, immunologists have already described many events of cross-reactivity between epitopes sharing less than 50% of the linear amino acid sequence, and this phenomenon has proven to be far more difficult to predict.

Considering that CTL stimulation will be triggered by the specific recognition of a pMHC structure by a given TCR, our group analyzed the TCR-interacting surface of 28 pMHC complexes presenting natural-occurring variants of an important HCV-derived epitope. These peptides were previously tested i*n vitro* by Fytili et al. 2008 [13], presenting great variability against a specific CTL population. We focused our structural analysis mainly on topology and charge distribution (obtained with GRASP2 program), and our results indicated the sharing of these features among complexes that stimulated response *in vitro*. Furthermore, the most remarkable difference in electrostatic potential was seen precisely in the only complex that presented no *in vitro* stimulation (Figure 10). In order to quantify this relations, we extract the color histogram of a selected area in the pMHC surface, and used as input for a Principal Component Analysis (PCA). Simple scatter plot of the 28 complexes according two extracted principal components clustered the pMHCs exactly in the same way that was observed by *in vitro* experiments. Since this approach was able to separate variants with reduced immunogenicity, it has a potential use for predicting the impact of viral scape mutations over the CTL recognition.

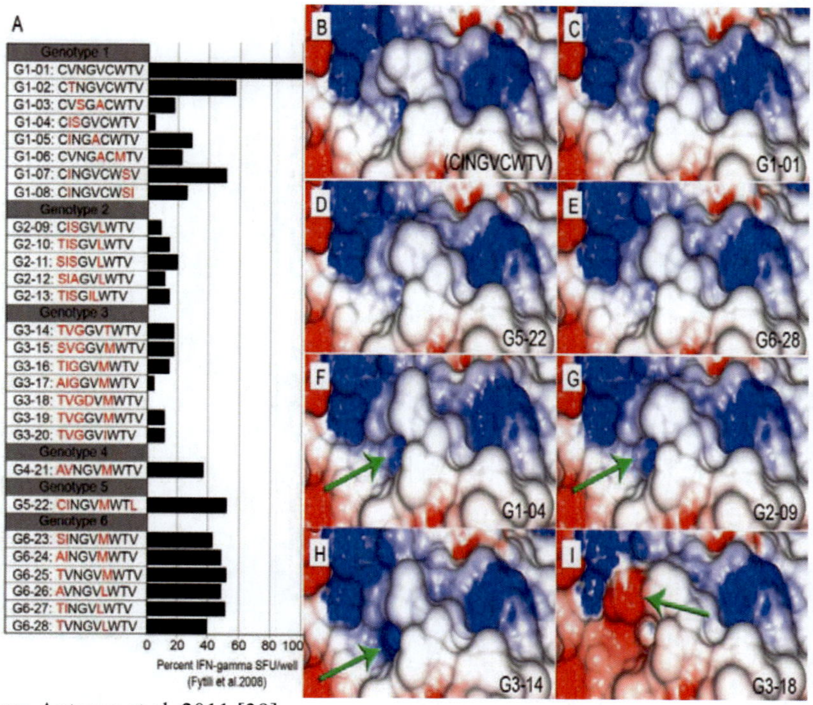

Modified from Antunes et al. 2011 [39].

Figure 10. Structural analysis of cross-reactivity. (A) Results of an *in vitro* assay performed by Fytili et al. 2008 [13]. Twenty eight $NS3_{1073}$ naturally occurring variants were tested for IFN-gamma production against a $NS3_{1073}$ wild-type-specific CD8+ T-cell population. (B-I) TCR-interacting area of pMHC complexes presenting some of these peptides. Charged areas over the complexes surface were computed with GRASP2 program and represented in a range from −10 kiloteslas (red spots) to +10 kiloteslas (blue spots). Green arrows indicate major differences in the electrostatic potential. Complexes with strong *in vitro* response (B-E) presented similar topology and charge distribution, while those with poor *in vitro* response (F-I) presented charged areas that differentiate them from the wild-type complex.

In the same work, we used a similar approach to perform a virtual screening of cross-reactive targets among 55 complexes available at the CrossTope Data Bank, based on information extracted from seven selected regions over the TCR-interacting surface [39]. In this chapter, we will present and discuss an updated version of this analysis. Sixty unrelated pMHC complexes were selected from the CrossTope DataBank for Cross-Reactivity Assessment, being 5 crystal structures and 55 *in silico* predicted structures (obtained with the previously described approach *D1-EM-D2*). They presented virus-derived peptides in the context of the most frequent human MHC allotype (HLA-A*02:01). Images of the TCR-interacting surface of these complexes, presenting the electrostatic potential distribution, were used to extract the color histograms (RGB) of seven selected areas, corresponding to the points of interaction with already described public TCRs. Values of mean and standard deviation of the three RGB components, for each one of these selected areas, were used as input for multivariate statistical methods, aiming to predict possible targets of cross-reactivity.

This selected dataset included some peptides with already known cross-reactivity, and the result of a Hierarchical Cluster Analysis (HCA) of these data agreed with this experimental background (Figure 11).

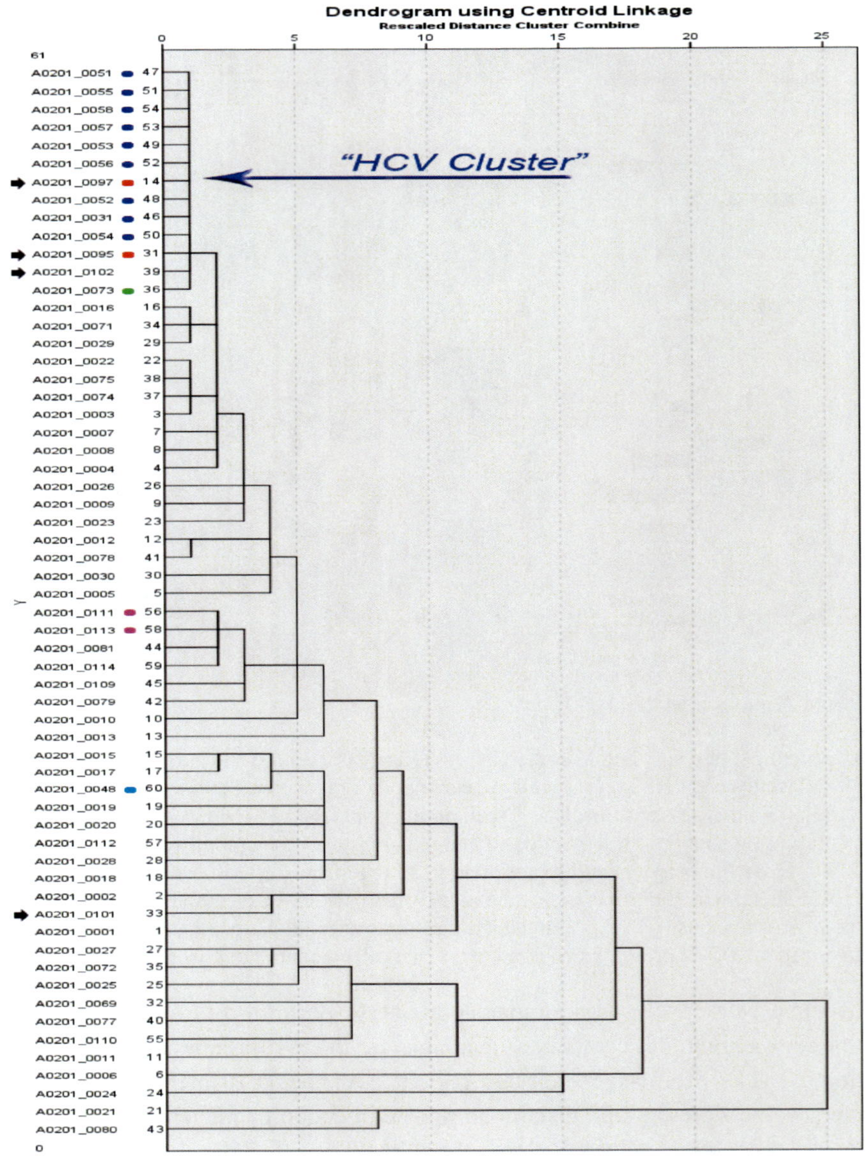

Figure 11. Dendrogram of a Hierarchical Cluster Analysis (HCA) with pMHC complexes. Sixty pMHC complex from CrossTope DB were clustered with a HCA, assisted by the SPSS software (PASW Statistics 18, IBM, Chicago IL. USA). CrossTope ID of each complex is provided on the left. Nine cross-reactive variants of the wild-type epitope HCV-NS3$_{1073}$ were placed in the same cluster (blue circles) while a noncross-reactive variant (cyan circle) was placed in a completely unrelated group. Other well-known cases of cross-recognition between Variola/Vaccinia epitopes (pink circles), and between IV/HIV epitopes (red circles), were also represented. Interestingly, these last two epitopes were also placed in the "HCV cluster". This same HIV epitope was already described as being cross-reactive with the HCV-NS5b peptide, in HIV/HCV co-infected individuals. A potential cross-reactive target of Human herpesvirus 4 (green circle) was also indicated. Crystal structures are indicated with black arrows. The remaining structures were predicted with the previously described *D1-EM-D2* approach.

For instance, we included 10 variants of the wild-type epitope HCV-NS3$_{1073-1081}$ (CV/INGVCWTV). The wild-type and all the cross-reactive variants (genotypes 4, 5 and 6) were placed in the same group, while the non cross-reactive variant (genotype 3) was placed in a completely unrelated group. The HCA dendrogram also clustered two cross-reactive variants of the epitope A26L6-14, one from *Vaccinia* and other from *Variola virus*. Other well-known case of cross-recognition, between the epitopes IV-M$_{158-66}$ (GILGFVFTL) and HIV-GAG$_{77-85}$ (SLYNTIAVL), was also represented, with both complexes placed in the same group. Interestingly, these last two epitopes were placed exactly in the same group of the cross-reactive variants of HCV-NS3$_{1073-1081}$ epitope.

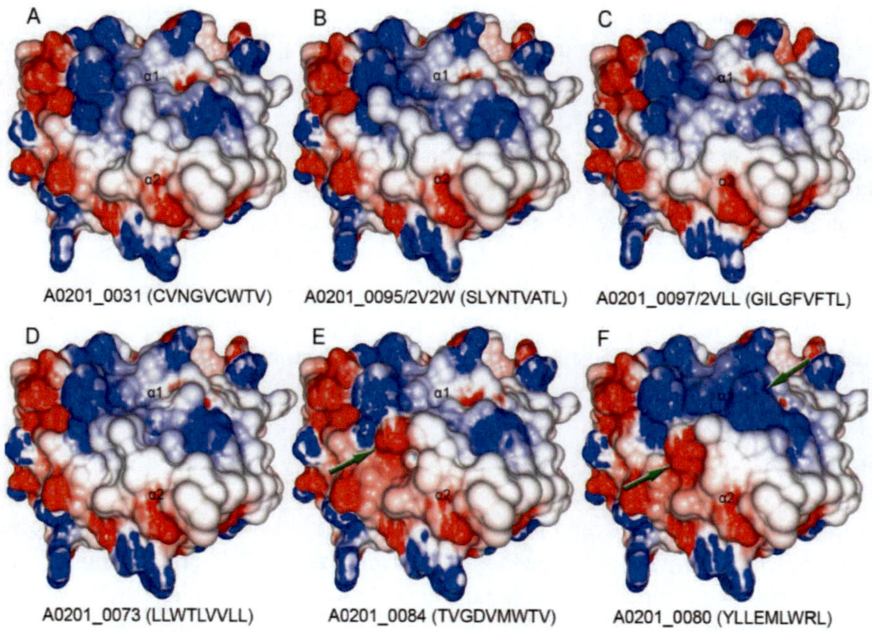

Figure 12. TCR-interacting surface of selected complexes. Images of the sixty pMHC complexes were obtained with GRAP2 software and used to extract the information for the HCA. Epitopes are placed between α1 and α2 domains of the MHC. "A" presents an HCV immunodominant epitope and "D" presents a non cross-reactive escape variant of the same virus. "B" and "C" are well-known cross-reactive targets. Complexes presenting similar surfaces (such as A, B, C and D) were grouped in the same cluster. Complexes presenting divergent surfaces (such as E and F) were placed in different groups. Green arrows indicate major differences in the electrostatic potential. PDB code of the two crystal structures is also presented.

As discussed in the original work [39], cross-reactivity between this HCV immunodominant epitope and the HIV-GAG$_{77-85}$ peptide has not been described so far. However, this same HIV epitope was already described as being cross-reactive with the HCV-NS5b peptide, in HIV/HCV co-infected individuals. There is yet another complex included in the same cluster, presenting the "LLWTLVVLL" peptide, from the Human herpesvirus 4 (LMP2$_{329-337}$). It is important to note that this peptide does not share even a single amino acid with the target peptide (CV/INGVCWTV) and, nevertheless, presented almost the same structural pattern when presented in the context of HLA-A*02:01 (Figure

12). This result stresses the power of this structural approach in prospecting new cross-reactive targets.

Assessing pMHC Stability through Molecular Dynamics

The first electric programmable program was invented in the 40's, while the DNA structure discovery occurred in the early 1950's. However, these two areas (computation and biology) find each other only in the 90's, with the great expansion of genome sequencing data of several organisms. At this moment, the need for computational "*know-how*" to store, share and analyze the huge generated sequencing data, forced biologists to include a whole IT staff and resources as an important part of the genome sequence programs. Although theoretical chemists at have been applying computational resources to model biochemical structures since the early 70's (in a field called molecular modeling), it is said that the genomic age led to the emergence of a new knowledge field: the Bioinformatics.

Initially, the bioinformatics was used only to analyze and help to get conclusions about genomes sequences. Later, it was also used over protein sequences. Because of this, sequence-based bioinformatics studies are referred as classic bioinformatics. However, new experimental techniques have allowed elucidating the DNA structure and the protein structure as well, such as X-ray crystallography and nuclear magnetic resonance (NMR). To study molecular structural composition, another bioinformatics subarea called structural bioinformatics has arisen.

Since immunology is the topic of this chapter, we will focus our interests in the study through structural bioinformatics to perform molecular simulations of pMHC complexes. One of the first molecular dynamics studies was performed by Oldfield and Chapman, in the 70's. They actually use nuclear magnetic resonance spectroscopy (NMR) and differential scanning spectroscopy to study the molecular dynamics of cerebroside-cholesterol and sphingomyelin-cholesterol interactions [45]. Although this had been a molecular dynamics study, the data were assayed strictly *in vitro*, not *in silico*.

In the 90's, Rognan et al. performed molecular dynamics simulation with HLA-B*27:05, anchored to different peptides, which were demonstrated, through *in vitro* studies, able to stimulate differential immunological responses. The objective of this study was to evidence that the molecular dynamics could be able to differentiate epitopes with different immunogenic properties. They could observe some differences, but the time simulation at the time was very low, compared to more recent studies, which could prevent the observation of some important aspects, such as secondary protein structure formation/deformation [46].

More recently, it had become possible to improve molecular dynamics systems using large molecules, as well relatively large time simulations (upon microseconds). Wan et al., for example, simulated a more complete pMHC/TCR interaction, with a model comprising MHC, TCR, peptide, CD4 and two opposing sections of a fosfolipid membrane along 10ns (the time needed to simulate a nanosecond on such a system was about 23.00 h) [47].

The use of pMHC molecular dynamics could be very useful to study cases beyond the static view of crystallography. It allows the calculation of molecules free energy, as well the computation of molecular phenomena that otherwise could not be seen [48]. For example, it is known that protein loops are very difficult to be seen through crystallography. Besides, their position and behavior in aqueous solution is not the same as in a crystallography unit. In this way, the molecular dynamics plays an important role to evidence these differences according to the environment where the molecule is found.

As already mentioned, in a previous publication we postulate that the immunogenicity of a given pMHC will be determined by three selection levels: Processing, Stability and Reactivity. While the first level can be accurately predicted by sequence-based bioinformatics tools, the later levels require the involvement of structure-based analysis. For Stability assessment, molecular dynamics stands out as the most appropriate tool.

In vitro studies have already associated pMHC stability with T Cell stimulation levels [49]. *In silico* simulations of pMHC dynamics, can provide useful information about a given pMHC complex stability, allowing to infer if this specific pMHC will be able to remain at the cell surface – structurally intact – for time enough to be recognized by a specific TCR.

Molecular dynamics can also be used as part of a larger analysis, to infer Reactivity properties of a given pMHC. In a recent work, our group analyzed 28 naturally occurring HCV-$NS3_{1073}$ variants in the context of HLA-A*02:01 allotype [39]. Structural analysis of the TCR-interacting surfaces of this complex indicated the sharing of structural features (topology and electrostatic potential) among cross-reactive complexes (previously tested *in vitro*). A short position-restrained molecular dynamics was carried out with all complexes in order to perform a full solvation of the structures, gradually heating the system up to 310 K. Structural analysis taken after this short simulation presented results in agreement with our first observations. More important, molecular dynamics also seemed to smooth out the electrostatic potential over the pMHC complex, possibly affecting a Principal Component Analysis that was used to group the cross-reactive complexes.

In a recent work, Stavrakoudis et al. (2011) [50] performed a molecular dynamics simulation to access the stability of the modeled MUC1-9/H2-K^b complex. They performed a complete RMSD and RMSF analysis of two independent trajectories, using as a control a third trajectory from a H2-K^b reference structure SEV9/H2-K^b complex (PDB code 1KPV), simulated under the same conditions. They also evaluated the hydrogen bond interactions between SEV9 and MUC1-9 peptides and the H-2K^b molecule, during the entire simulation. In addition, the authors also calculated the Buried Surface Area (BSA), using the NACCESS program (http://www.bioinf.manchester. ac.uk/naccess/)[22]. The Buried Surface Area is referred as a good indicator of the binding of a ligand into a protein [51] and, in this case, BSA of the peptide/H-2K^b interface remained constant during the simulation, which indicates complex stability. Based on their results, the authors conclude that a stable complex is formed and that MHC binding pockets A and F interacted closely with both N and C-terminus of the peptide, which played an important role in stabilizing the complex [50].

Conclusion

The understanding of some immunological mechanisms still remains as a major challenge in modern science, particularly regarding the elements involved in immunogenicity. This great interest arises on the basis that such knowledge can be applied directly to areas of critical importance, such as vaccines development and immunotherapies against cancer.

Bioinformatics emerged as a tool that can assist to solve this kind of problems, mainly because of technical limitations to deal with a very huge number of elements (vaccine and therapeutic targets). The automated analysis of large-scale elements looking for patterns that are important for discriminating between self/non-self antigens, or antigens that are differentially expressed in tumors, will allow disentangling the issues that are still nebulous in this area of immunology. Especially because the genetic diversity of populations and pathogens has often been underestimated in classical studies of vaccinology, which may explain why many vaccines do not have the expected spectrum of activity. However, we cannot restrict the analysis of proteins and peptides involved in these processes solely to sequence analysis, or once again we will be neglecting crucial factors in the full understanding of the immunogenicity. Thus structural immunoinformatics emerges as a field with great potential to fill these gaps. The growing number of structural databases hosting immune system proteins, and the different strategies to obtain and analyze pMHC complexes, as previously described in this chapter, open the way for a new age of research. These innovative approaches allow the development of more rationalized vaccines and customized therapies, central factors for the eradication of infectious diseases that have long plague mankind.

References

[1] Theves, G. [Smallpox: an historical review]. Bulletin de la Société des sciences médicales du Grand-Duché de Luxembourg, 1997 134, 31-51.

[2] Lecoq, H. [Discovery of the first virus, the tobacco mosaic virus: 1892 or 1898?]. Comptes rendus de l'Académie des sciences Série III, Sciences de la vie, 2001 324, 929-33.

[3] Fischer Lindahl, K. On naming H2 haplotypes: functional significance of MHC class Ib alleles. *Immunogenetics*, 1997 46(1), 53-62.

[4] Tonegawa, S. Somatic generation of antibody diversity. *Nature*, 1983 302, 575-81.

[5] Tonegawa, S; Steinberg, C; Dube, S; Bernardini, A. Evidence for somatic generation of antibody diversity. *Proceedings of the National Academy of Sciences of the United States of America,* 1974 71, 4027-31.

[6] Vieira, GF; Chies, JA. Immunodominant viral peptides as determinants of cross-reactivity in the immune system--Can we develop wide spectrum viral vaccines? *Med. Hypotheses*, 2005 65(5), 873-9.

[7] Tong, JC; Ren, EC. Immunoinformatics: Current trends and future directions. *Drug Discov. Today,* 2009 Jul 14(13-14), 684-9.

[8] Korber, B; LaBute, M; Yusim, K. Immunoinformatics comes of age. *PLoS Comput. Biol*, 2006 Jun 30 2(6), e71.

[9] Tomar, N; De, RK. Immunoinformatics: an integrated scenario. *Immunology*, 2010, 153-68.

[10] Rappuoli, R. Reverse vaccinology. *Curr. Opin. Microbiol*, 2000 Oct 3(5), 445-50.

[11] Vivona, S; Gardy, JL; Ramachandran, S; Brinkman, FSL; Raghava, GPS; Flower, DR, et al. Computer-aided biotechnology: from immuno-informatics to reverse vaccinology. *Trends in biotechnology*, 2008 26, 190-200.

[12] Frankild, S; de Boer, RJ; Lund, O; Nielsen, M; Kesmir, C. Amino acid similarity accounts for T cell cross-reactivity and for "holes" in the T cell repertoire. *PLoS ONE,* 2008 3(3), e1831.

[13] Fytili, P; Dalekos, GN; Schlaphoff, V; Suneetha, PV; Sarrazin, C; Zauner, W, et al. Cross-genotype-reactivity of the immunodominant HCV CD8 T-cell epitope NS3-1073. *Vaccine*, 2008 Jul 23 26(31), 3818-26.

[14] Pettersen, EF; Goddard, TD; Huang, CC; Couch, GS; Greenblatt, DM; Meng, EC, et al. UCSF Chimera--a visualization system for exploratory research and analysis. *J. Comput. Chem*, 2004 Oct 25(13), 1605-12.

[15] Petrey, D; Honig, B. GRASP2: visualization, surface properties, and electrostatics of macromolecular structures and sequences. *Methods Enzymol*, 2003 374, 492-509.

[16] Coombes, JL; Powrie, F. Dendritic cells in intestinal immune regulation. *Nat. Rev. Immunol*, 2008 Jun 8(6), 435-46.

[17] Vita, R; Zarebski, L; Greenbaum, JA; Emami, H; Hoof, I; Salimi, N, et al. The immune epitope database 2.0. *Nucleic Acids Res*, 2010 Jan 38(Database issue), D854-62.

[18] Rigo, MM; Antunes, DA; Sinigaglia, M; Fülber, CC; Chies, JAB; Vieira, GF. Molecular Aspects Involved in the Immunogenicity against Viral Epitopes : An Immunoinformatic Perspective. In: Villanueva CJ, editor. *Immunogenicity*. Nova York , NY: Nova Publishers; 2011. p. 1-23.

[19] Giudicelli, V; Chaume, D; Bodmer, J; Muller, W; Busin, C; Marsh, S, et al. IMGT, the international ImMunoGeneTics database. Nucleic Acids Res, 1997 Jan 1 25(1), 206-11.

[20] Lefranc, MP; Giudicelli, V; Kaas, Q; Duprat, E; Jabado-Michaloud, J; Scaviner, D, et al. IMGT, the international ImMunoGeneTics information system. *Nucleic Acids Res,* 2005 Jan 1 33(Database issue), D593-7.

[21] McDonald, IK; Thornton, JM. Satisfying hydrogen bonding potential in proteins. *J. Mol. Biol*, 1994 May 20 238(5), 777-93.

[22] Hubbard, SJ; Thornton, JM. NACCESS Computer Program. Department of Biochemistry and Molecular Biology, University College London, 1993.

[23] Laskowski, RA. SURFNET: a program for visualizing molecular surfaces, cavities, and intermolecular interactions. *J. Mol. Graph*, 1995 Oct 13(5), 323-30, 07-8.

[24] Wallace, AC; Laskowski, RA; Thornton, JM. LIGPLOT: a program to generate schematic diagrams of protein-ligand interactions. *Protein Eng*, 1995 Feb 8(2), 127-34.

[25] Crooks, GE; Hon, G; Chandonia, JM; Brenner, SE. WebLogo: a sequence logo generator. *Genome Res*, 2004 Jun 14(6), 1188-90.

[26] Antunes, DA; Vieira, GF; Rigo, MM; Cibulski, SP; Sinigaglia, M; Chies, JA. Structural allele-specific patterns adopted by epitopes in the MHC-I cleft and reconstruction of MHC:peptide complexes to cross-reactivity assessment. *PLoS ONE*, 2010 5(4), e10353.

[27] Parham, P. Functions for MHC class I carbohydrates inside and outside the cell. *Trends in biochemical sciences*, 1996 21, 427-33.

[28] Horton, R; Wilming, L; Rand, V; Lovering, RC; Bruford, Ea; Khodiyar, VK, et al. Gene map of the extended human MHC. *Nature reviews Genetics*, 2004 5, 889-99.

[29] Vandiedonck, C; Knight, JC. The human Major Histocompatibility Complex as a paradigm in genomics research. *Briefings in functional genomics and proteomics*, 2009 8, 379-94.

[30] Cardozo, T; Totrov, M; Abagyan, R. Homology modeling by the ICM method. *Proteins*, 1995 23, 403-14.

[31] Bordner, AJ; Abagyan, R. Ab initio prediction of peptide-MHC binding geometry for diverse class I MHC allotypes. *Proteins*, 2006 May 15 63(3), 512-26.

[32] Khan, JM; Ranganathan, S. pDOCK: a new technique for rapid and accurate docking of peptide ligands to Major Histocompatibility Complexes. *Immunome research*, 2010 6 Suppl 1, S2.

[33] Khan, JM; Ranganathan, S. A multi-species comparative structural bioinformatics analysis of inherited mutations in alpha-D-mannosidase reveals strong genotype-phenotype correlation. *BMC genomics*, 2009 10 Suppl 3, S33.

[34] Cárdenas, C; Bidon-Chanal, A; Conejeros, P; Arenas, G; Marshall, S; Luque, FJ. Molecular modeling of class I and II alleles of the major histocompatibility complex in Salmo salar. *Journal of computer-aided molecular design*, 2010 24, 1035-51.

[35] Rammensee, H; Bachmann, J; Emmerich, NP; Bachor, OA; Stevanovic, S. SYFPEITHI: database for MHC ligands and peptide motifs. *Immunogenetics*, 1999 Nov 50(3-4), 213-9.

[36] Tian, F; Lv, F; Zhou, P; Yang, Q; Jalbout, AF. Toward prediction of binding affinities between the MHC protein and its peptide ligands using quantitative structure-affinity relationship approach. *Protein Pept. Lett*, 2008 15(10), 1033-43.

[37] Meijers, R; Lai, CC; Yang, Y; Liu, JH; Zhong, W; Wang, JH, et al. Crystal structures of murine MHC Class I H-2 D(b) and K(b) molecules in complex with CTL epitopes from influenza A virus: implications for TCR repertoire selection and immunodominance. *J. Mol. Biol,* 2005 Feb 4 345(5), 1099-110.

[38] Fülber, CC; Antunes, DA; Rigo, MM; Artur, J; Chies, B; Sinigaglia, M, et al., editors. Reconstruction of MHC Alleles by Cross Modeling and Structural Assessment. BIOCOMP'10; 2010; Las Vegas, USA.

[39] Antunes, DA; Rigo, MM; Silva, JP; Cibulski, SP; Sinigaglia, M; Chies, JAB, et al. Structural *in silico* analysis of cross-genotype-reactivity among naturally occurring HCV NS3-1073-variants in the context of HLA-A*02:01 allele. *Molecular immunology*, 2011.

[40] Trott, O; Olson, AJ. AutoDock Vina: Improving the speed and accuracy of docking with a new scoring function, efficient optimization, and multithreading. *J. Comput. Chem,* 2009 Jun 4.

[41] Van Der Spoel, D; Lindahl, E; Hess, B; Groenhof, G; Mark, AE; Berendsen, HJ. GROMACS: fast, flexible, and free. *J. Comput. Chem*, 2005 Dec 26(16), 1701-18.

[42] Arnautova, YA; Abagyan, RA; Totrov, M. Development of a new physics-based internal coordinate mechanics force field and its application to protein loop modeling. *Proteins*, 2011 79, 477-98.

[43] Seifert, MHJ; Kraus, J; Kramer, B. Virtual high-throughput screening of molecular databases. Current opinion in drug discovery and development, 2007 10, 298-307.

[44] Davis-Harrison, RL; Insaidoo, FK; Baker, BM. T cell receptor binding transition states and recognition of peptide/MHC. *Biochemistry*, 2007 46, 1840-50.

[45] Oldfield, E; Chapman, D. Molecular dynamics of cerebroside-cholesterol and sphingomyelin-cholesterol interactions: Implications for myelin membrane structure. *FEBS Lett,* 1972 Apr 1 21(3), 303-6.

[46] Rognan, D; Scapozza, L; Folkers, G; Daser, A. Molecular dynamics simulation of MHC-peptide complexes as a tool for predicting potential T cell epitopes. *Biochemistry*, 1994 Sep 27 33(38), 11476-85.

[47] Wan, S; Flower, DR; Coveney, PV. Toward an atomistic understanding of the immune synapse: large-scale molecular dynamics simulation of a membrane-embedded TCR-pMHC-CD4 complex. *Mol. Immunol*, 2008 Mar 45(5), 1221-30.

[48] Wan, S; Coveney, PV; Flower, DR. Molecular basis of peptide recognition by the TCR: affinity differences calculated using large scale computing. *J. Immunol*, 2005 Aug 1 175(3), 1715-23.

[49] van der Burg, SH; Visseren, MJ; Brandt, RM; Kast, WM; Melief, CJ. Immunogenicity of peptides bound to MHC class I molecules depends on the MHC-peptide complex stability. *Journal of immunology* (Baltimore, Md : 1950), 1996 156, 3308-14.

[50] Stavrakoudis, A; Tsoulos, IG; Uray, K; Hudecz, F; Apostolopoulos, V. Homology modeling and molecular dynamics simulations of MUC1-9/H-2K(b) complex suggest novel binding interactions. *Journal of molecular modeling,* 2011 17, 1817-29.

[51] Olsson, TS; Williams, MA; Pitt, WR; Ladbury, JE. The thermodynamics of protein-ligand interaction and solvation: insights for ligand design. *J. Mol. Biol*, 2008 Dec 26 384(4), 1002-17.

In: Bioinformatics Research: New Developments
Editors: C. Battik and K. Belhassine
ISBN: 978-1-61942-363-3

Chapter II

Natural Selection, Molecular Recombination Structure and Evolutionary Historical Relationships among Members of the Subfamily *Comovirinae* (Family: *Secoviridae*): Towards a Possible Reclassification

Moncef Boulila*
Institut de l'Olivier, Sousse Ibn-Khaldoun, Tunisia

Abstract

In order to determine putative recombination events and their frequency in the RNAs 1 and 2 of currently the known members of the subfamily *Comovirinae*, a detailed study comprising 107 accessions retrieved from the international databases, was carried out by using RECCO and RDP v3.31β softwares. These programs allowed the detection of potential recombination sites in all the three virus genera composing this subfamily with various degrees of consistency. Furthermore, based on Fisher's Exact Test of Neutrality and contrary to RNA 1, positive selection exerted on protein-coding genes was detected in the RNA 2 of numerous representatives included in the subfamily *Comovirinae*. The reconstructed trees showed that *Raspberry ringspot virus* constitutes an independent cluster clearly separated from members of the current Subgroup A of *Nepovirus* genus. In addition, *Grapevine deformation* and *Grapevine Anatolian ringspot* viruses should be included in the subgroups A and B of the same genus, respectively, instead of being considered as two unclassified nepoviruses.

* Corresponding author: Moncef Boulila, Institut de l'Olivier, B.P. 14, 4061 Sousse Ibn-Khaldoun, Tunisia. phone/fax: 00216 73236135, E-mail: boulila.moncef@ yahoo.fr.

Keywords: Bioinformatics, phylogeny, recombination, selection, *Comovirinae*, Sequence, Taxonomy

Introduction

Recombination is a fundamental biological process that can increase viral pathogenicity by diffusing genetic material throughout populations (Awadalla 2003). The biological mechanisms of recombination differ across organisms, but in broad terms recombination results in the creation of mosaic sequences where the evolutionary history at each site may be different. Recombination, defined as the exchange of genetic information between two nucleotide sequences, is an important process that influences biological evolution at many different levels. Recombination explains a considerable amount of genetic diversity in natural populations and, in general, genes located in regions of the genome with low levels of recombination have low levels of polymorphism (Posada and Crandall 2001). Recombination reshuffles existing variation and even creates new variants. A single virus isolate does not consist of a single RNA sequence, but of a population of related sequence variants, often referred to as quasispecies (Holland et al. 1992; Domingo et al. 1995; Eigen 1996). The quasispecies nature of RNA viruses implies a high adaptive potential, allowing for the rapid selection of biologically distinct sequence variants with the highest fitness in new environments. It may result in dramatic changes in the biological properties of the virus, with major epidemiological consequences, including the appearance of resistance-breaking strains or the acquisition of broader host ranges (Legg and Thresh 2000; Monci et al. 2002). It has been shown that RNA recombination enables the exchange of genetic material not only between the same or similar viruses but also between distinctly different viruses (Worobey and Holmes 1999). Sometimes, it also permits crossovers between viral and host RNA (Greene and Allison 1994; Aaziz and Tepfer 1999, Baroth et al. 2000, Nagai et al. 2003). Taking into account the structure of viral genomic molecules and the location of crossover sites, three basic types of RNA recombination were distinguished: Homologous, aberrant homologous and non-homologous (Lai 1992; Alejska et al. 2001). The former two occur between two identical or similar RNAs (or between molecules displaying local homology), while the latter involves two different molecules. Most of the collected data suggest that RNA recombinants are formed according to a copy choice model (Alejska et al. 2001). A viral replication complex starts nascent RNA strand synthesis on one template, called RNA donor and then switches to another template, called RNA acceptor. Accordingly, two main factors are thought to affect RNA recombination: the structure of recombining molecules and the ability of the viral replicase to switch templates. Through generations, viral populations evolve under various selective forces at different regions and sites that display different functional constraints. A stringent and robust criterion for detecting adaptive evolution in a protein-coding gene is an accelerated nonsynonymous (d_N, amino acid replacing) rate relative to the synonymous (d_S, silent) rate of substitutions, with the rate ratio $\omega = d_N/d_S > 1$. As silent mutations do not change the amino acid whereas replacement mutations do, the difference in their fixation rates provides a measure of selective pressure on the protein.

Amongst positive-strand plant RNA viruses, the subfamily *Comovirinae* (family: *Secoviridae*) encompasses several viruses having an important economical impact. According to Sanfaçon et al. (2009), the subfamily *Comovirinae* consists of three genera of plant viruses with bipartite genome: *Comovirus* (type species: *Cowpea mosaic virus*), *Fabavirus* (type species: *Broad bean wilt virus* 1) and *Nepovirus* (type species: *Tobacco ringspot virus*). The genome consists of two species of linear positive sense single stranded RNA. Both RNAs are necessary for systemic infection. Sizes of RNAs differ among genera: *Nepovirus* RNA 1 (7.2-8.4 kb) and RNA 2 (3.9-7.2 kb) are larger than *Fabavirus* and *Comovirus* RNA 1 (5.9-7.2 kb) and RNA 2 (3.5-4.5 kb). The 5'-end of the genome has a genome-linked protein (VPg). The 3'-terminus has conserved nucleotide sequences. The 3'-terminus has a poly (A) tract. The multipartite genome is divided among different particles, each segment encapsidated separately and the segments are distributed between two different types of particles. The largest particles contain each one molecule of RNA 1 (sedimenting component B). The medium sized particles contain each one molecule of RNA 2 (sedimenting component M). The smallest particles contain no nucleic acid and are found in the top (sedimenting component T) band after sedimentation. It should be noted that some representatives of the *Nepovirus* genus possess a satellite RNA such as *Grapevine fanleaf virus* F 13 strain (Pinck et al. 1988; Fuchs et al. 1989).

Both recombination and selective constraints can influence the evolutionary relationships among different viruses (Massey et al. 2008; Boulila 2009b,c; 2010).

Revealing the evolutionary history of the members of RNA virus families has become one of the main goals for evolutionary virologists (Costa-Mattioli et al. 2002; Katayama et al. 2002; Lukashov and Goudsmit 2002; de Villiers et al. 2004; Thiery et al. 2004), and the proposals for classification have been changing constantly in recent years. Many causes of this change are: (i) the lack of complete genome sequences in databanks leading to focusing on the analysis of a single gene, (ii) the presence of extensive recombination, (iii) the poor reliability of alignments due to the typically high sequence divergence of viruses and the existence of deletions and insertions. However, new techniques have been recently developed for analyzing unaligned multiple whole genomes and for producing global phylogenies (Stuart et al. 2002a,b; Stuart and Berry 2003; Qi et al. 2004a,b).

The objective of the present study was to investigate and characterize putative recombination events and selective constraints in the complete genome of currently the known members of the subfamily *Comovirinae* and finally to propose a new classification based on the predicted evolutionary relationships among them. This was achieved by analyzing the whole genome of 107 accessions retrieved from GenBank.

Materials and Methods

Virus Sequence Source

The sequences of RNAs 1 (32 accessions) and 2 (75 accessions) used in this study were downloaded from GenBank (Table 1).

Sequence Alignments, Selective Pressure, Recombination and Phylogenetic Analyses

The nucleotide sequences were aligned using CLUSTALW 2.0.9 and CLUSTALX 2.0.9 (Larkin et al. 2007) with default parameters. Their phylogenetic relationships were determined by five methods using the neighbor-joining (NJ), Minimum Evolution (ME), Maximum Parsimony (MP), and Unweighted Pair Group Method with Arithmetic mean (UPGMA) implemented in MEGA4.1β program (Kumar et al. 2008) and Maximum-likelihood (ML) using PHYLO_WIN graphic tool (Galtier et al. 1996). Bootstrap analyses with 500 replicates were performed to assess the robustness of the branches.

Positive selection, using MEGA4.1β program, was inferred by the counting method described by Nei and Gojobori (1986) and later by Suzuki and Gojobori (1999). This method utilized the phylogenetic tree of sequences analyzed. In the parsimony method, the total numbers of synonymous (c_S) and nonsynonymous (c_N) substitutions as well as the average numbers of synonymous (s_S) and nonsynonymous (s_N) sites per codon over the phylogenetic tree for each codon site were computed according to the maximum parsimony principle (Fitch 1971; Hartigan 1973). The null hypothesis of selective neutrality (r_S=r_N or $\omega = 1$) was tested for each site by computing the probability (p) of obtaining the observed or more biased values for c_S and c_N, which were assumed to follow a binomial distribution with the probabilities of occurrence of synonymous and nonsynonymous substitutions given by $s_S/(s_S + s_N)$ and $s_N/(s_S + s_N)$, respectively. Positive selection is inferred when $p<0.05$ and $c_N/s_N>c_S/s_S$ (Suzuki, 2006).

Potential recombination events between diverged nucleotide sequences were explored with two programs: RDP v3.31β (Martin et al. 2005b) and RECCO (Maydt and Lengauer 2006). RDP incorporates several published recombination detection methods into a single suite of tools: RDP (Martin and Rybicki 2000), GENECONV (Padidam et al. 1999), BOOTSCAN (Martin et al. 2005a), MAXCHI (Smith 1992), CHIMAERA (Posada and Crandall 2001), SISCAN (Gibbs et al. 2000), and 3SEQ (Boni et al. 2007). In all cases, defaults parameters were used. Only events predicted by more than half of the methods are considered as significant. The algorithm developed and described by Maydt and Lengauer (2006) as fast, simple and sensitive method for detecting recombination in a set of sequences and locating putative recombination breakpoints is based on cost minimization. This method has only two tunable parameters, recombination and mutation cost. In practice the only parameter considered is α, representing the cost of mutation relative to recombination.

When α changes from 0 to 1, the cost of mutation weighted by α increases, and the cost for recombination weighted by 1- α decreases. In other words, the parameter α controls the ambiguity between mutation and recombination.

Results

Recombination Events during Evolution of the *Comovirinae*

Examination of the RECCO program output regarding the occurrence of recombination events in RNA 1 of the *Comovirinae*, revealed that three out of 14 comoviruses were putative recombinants (i.e. CPMV, CPSMV, RCMV.S). In contrast, BPMV.KY G-7, BPMV.IL-

Carbondale 1/Sg.1 and 2, BPMV.k-Hancock 1, BPMV.k-Hopkins 1, RaMV.California, RaMV.Japanese, SqMV.CH 99-211, SqMV.Y, SqMV.Kimble and SqMV.Arizona Z, did not show any recombinant signal (Table 2).

Table 1. Current taxonomic status of members of the subfamily *Comovirinae* included in the study and their accession numbers

Genus (in bold) and unclassified viruses	Virus/Isolate	GenBank accession number	
		RNA1	RNA2
Comovirus	Bean pod mottle virus/KY G-7 (BPMV/KY G-7)	NC_003496	NC_003495
	Bean pod mottle virus/IL-Carbondale 1-subgroup II (BPMV/IL-Carbondale 1/Sg.II)	AY744932	AY744933
	Bean pod mottle virus/K-Hanckock 1 (BPMV/K-Hanckock 1)	AF394606	AF394607
	Bean pod mottle virus/K-Hopkins 1 (BPMV/K-Hopkins 1)	AF394608	AF394609
	Bean pod mottle virus/IL-Carbondale 1-subgroup I (BPMV/IL-Carbondale 1/sg I)	AY744931	/
	Cowpea mosaic virus (CPMV)	NC_003549	NC_003550
	Cowpea severe mosaic virus (CPSMV)	NC_003545	NC_003544
	Radish mosaic virus/California (RaMV/California)	AB456531	AB456532
	Radish mosaic virus/Japanese (RaMV/Japanese)	NC_010709	NC_010710
	Red clover mottle virus/S (RCMV/S)	NC_003741	NC_003738
	Squash mosaic virus/CH 99-211 (SqMV/CH 99-211)	EU421059	EU421060
	Squash mosaic virus/Y (SqMV/Y)	NC_003799	NC_003800
	Squash mosaic virus/Kimble (SqMV/Kimble)	/	AF059533
	Squash mosaic virus/Arizona.Z (SqMV/Arizona.Z)	/	AF059532
Fabavirus	Broad bean wilt virus 1/Ben (BBWV 1/Ben)	AY781171	/
	Broad bean wilt virus 1/Singapore (BBWV 1/Singapore)	/	AF225955
	Broad bean wilt virus 1/ATCC PV132 (BBWV 1/ATCC PV 132)	NC_005289	NC_005290
	Broad bean wilt virus 2/ME (BBWV 2/ME)	NC_003003	NC_003004
	Broad bean wilt virus 2/IA (BBWV 2/IA)	/	AB032403
	Broad bean wilt virus 2/B 935 (BBWV 2/B 935)	AF149425	/
	Broad bean wilt virus 2/MB7 (BBWV 2/MB7)	AB013615	AB013616
	Patchouli mild mosaic virus/Philippines (PMMV/Philippines)	NC_003975	NC_003974
Unclassified F. →	Gentian mosaic virus /N-1 (GMV/N-1)	AB084452	AB084453

Table 1. (Continued)

Genus (in bold) and unclassified viruses	Virus/Isolate	GenBank accession number	
		RNA1	RNA2
Nepovirus	Arabis mosaic virus/NW (ArMV/NW)	NC_006057	NC_006056
	Arabis mosaic virus/Lv (ArMV/Lv)	EU617326	EU617327
	Arabis mosaic virus/Lilly (ArMV/Lilly)	/	AB279741
	Arabis mosaic virus/Narcissus (ArMV/Narcissus)	/	AB279740
	Arabis mosaic virus/Butterbur (ArMV/Butterbur)	/	AB279739
	Tobacco ringspot virus/Bud blight (TRSV/Bud blight)	NC_005097	NC_005096
	Raspberry ringspot virus/Cherry (RpRSV/Cherry)	NC_005266	NC_005267
	Raspberry ringspot virus (RpRSV)	/	S46011
	Raspberry ringspot virus/grapevine (RpRSV/grapevine)	AY310444	AY310445
	Raspberry ringspot virus/RAC815 (RpRSV/RAC815)	/	EF534293
	Grapevine fanleaf virus/F 13 (GFLV/F 13)	NC_003615	NC_003623
	Grapevine fanleaf virus/Vol57c6 (GFLV/Vol57c6)	/	DQ922679
	Grapevine fanleaf virus/Vol57c5 (GFLV/Vol57c5)	/	DQ922678
	Grapevine fanleaf virus/Vol57c4 (GFLV/Vol57c4)	/	DQ922677
	Grapevine fanleaf virus/Vol57c3 (GFLV/Vol57c3)	/	DQ922676
	Grapevine fanleaf virus/Vol57c2 (GFLV/Vol57c2)	/	DQ922675
Subgroup A	Grapevine fanleaf virus/Vol57c1 (GFLV/Vol57c1)	/	DQ922674
	Grapevine fanleaf virus/Vol55c3 (GFLV/Vol55c3)	/	DQ922673
	Grapevine fanleaf virus/Vol55c2 (GFLV/Vol55c2)	/	DQ922672
	Grapevine fanleaf virus/Vol55c1 (GFLV/Vol55c1)	/	DQ922671
	Grapevine fanleaf virus/Vol54c3 (GFLV/Vol54c3)	/	DQ922670
	Grapevine fanleaf virus/Vol54c2 (GFLV/Vol54c2)	/	DQ922669
	Grapevine fanleaf virus/Vol54c1 (GFLV/Vol54c1)	/	DQ922668
	Grapevine fanleaf virus/Vol52c1 (GFLV/Vol52c1)	/	DQ922667
	Grapevine fanleaf virus/Vol51c5 (GFLV/Vol51c5)	/	DQ922666
	Grapevine fanleaf virus/Vol51c4 (GFLV/Vol51c4)	/	DQ922665
	Grapevine fanleaf virus/Vol51c3 (GFLV/Vol51c3)	/	DQ922664
	Grapevine fanleaf virus/Vol51c2 (GFLV/Vol51c2)	/	DQ922663
	Grapevine fanleaf virus/Vol51c1 (GFLV/Vol51c1)	/	DQ922662
	Grapevine fanleaf virus/Vol50c2 (GFLV/Vol50c2)	/	DQ922661
	Grapevine fanleaf virus/Vol50c1 (GFLV/Vol50c1)	/	DQ922660
	Grapevine fanleaf virus/Vol49c2 (GFLV/Vol49c2)	/	DQ922659
	Grapevine fanleaf virus/Vol49c1 (GFLV/Vol49c1)	/	DQ922658
	Grapevine fanleaf virus/Vol47c5 (GFLV/Vol47c5)	/	DQ922657

Genus (in bold) and unclassified viruses	Virus/Isolate	GenBank accession number	
		RNA1	RNA2
Nepovirus	Grapevine fanleaf virus/Vol47c4 (GFLV/Vol47c4)	/	DQ922656
	Grapevine fanleaf virus/Vol47c3 (GFLV/Vol47c3)	/	DQ922655
	Grapevine fanleaf virus/Vol47c2 (GFLV/Vol47c2)	/	DQ922654
	Grapevine fanleaf virus/Vol47c1 (GFLV/Vol47c1)	/	DQ922653
Subgroup A	Grapevine fanleaf virus/Vol45c1 (GFLV/Vol45c1)	/	DQ922652
	Grapevine fanleaf virus/B19a (GFLV/B19a)	/	AY780903
	Grapevine fanleaf virus/A10a (GFLV/A10a)	/	AY780902
	Grapevine fanleaf virus/A17d (GFLV/A17d)	/	AY780901
	Grapevine fanleaf virus/A17b (GFLV/A17b)	/	AY780900
	Grapevine fanleaf virus/A17a (GFLV/A17a)	/	AY780899
	Grapevine fanleaf virus/NW (GFLV/NW)	/	AY017338
Subgroup B	Beet ringspot virus/S (BRSV/S)	NC_003693	NC_003694
	Cycas necrotic stunt virus (CNSV)	NC_003791	NC_003792
	Grapevine chrome mosaic virus (GCMV)	NC_003622	NC_003621
	Olive latent ringspot virus (OLRSV)	/	AJ277435
	Tomato black ring virus/MJ (TBRV/MJ)	NC_004439	NC_004440
	Tomato black ring virus/ED (TBRV/ED)	/	X80831
Subgroup C	Blackcurrant reversion virus (BCRV)	NC_003509	NC_003502
	Peach rosette mosaic virus/Michigan grapevine (PRMV/Michigan grapevine) (PRMV/Michigan grapevine)	AF016626	/
	Tomato ringspot virus/Raspberry (ToRSV/Raspberry)	NC_003840	NC_003839
Unclassified nepoviruses	Grapevine Anatolian ringspot virus (GARSV)	/	AY291207
	Grapevine deformation virus (GDV)	/	AY291208

Table 2. Determination of inferred putative recombination events and their frequency along the sequences of RNA 1 of members of the subfamily *Comovirinae*

Recombination determined by RECCO								Recombination determined by RDP v3.31β
Virus.isolate			Length of breakpoint				Genomic position of the longest breakpoint (size in nucleotide)	putative parentals (Major x Minor)
	NRS	GIRE (nt)	1 residue	2 residues	3 residues	>3residues		
CPMV	79	1017-7527	25	14	12	28	7054-7069 (16) 7249-7264 (16)	/
CPSMV	97	916-7474	33	21	14	29	5102-5119 (18)	/
RCMV.S	78	1017-7528	29	16	9	24	3006-3022 (17)	/
BBWV1.ATCC PV132	2	2860-2951	0	1	0	1	2946-2951 (6)	/
BBWV2.ME	21	1493-6228	6	5	1	9	5550-5591 (42)	
BBWV2.B935	3	2238-3002	0	1	1	1	2985-3002 (8)	/
GMV.N-1	74	1131-7519	26	9	9	30	3437-3456 (20)	/
ArMV.NW	9	68-7869	8	0	0	1	3730-3737 (8)	TRSV.BB x BPMV. k-H1
ArMV.Lv	8	68-7869	6	0	2	0	3491-3493 (3) 3735-3737 (3)	/
TRSV.Bud blight (TRSV.BB)	188	917-7534	69	48	26	45	2144-2202 (59)	ArMV.NW x RaMV.jap. TBRV.MJ x BPMV. k-H1
RpRSV.Cherry	3	206-7760	3	0	0	0	206; 7601;7760 (1)	/
RpRSV.grapevine	3	206-7760	3	0	0	0	206;7601,7760 (1)	/
GFLV.F13	29	830-8001	8	3	6	12	5836-5855 (20)	TRSV.BB x BPMV. K-H1
BRSV.S	17	219-7966	15	1	0	1	2234-2237 (4)	TRSV.BB x BPMV. K-H1
CNSV	79	917-8056	32	13	8	26	4905-4926 (22)	TRSV.BB x BPMV. K-H1
GCMV	34	905-7532	16	6	7	5	4597-4655 (59)	TRSV.BB x BPMV K-H1
TBRV.MJ	15	154-7966	15	0	0	0	154;219;220;224;281; 719;734;925;2141;2237; 7536;7775;7787;7916; 7966 (1)	BRSV.S x GCMV TRSV.BB x BPMV K-H1
BCRV	215	916-7521	95	42	29	49	5480-5501 (22)	/
PRMV.Michigan grapevine	170	917-7518	74	32	18	46	5184-5202 (19)	/
ToRSV.Raspberry	176	916-7531	82	36	23	35	3596-3619 (24)	/

With RDP v3.31β algorithm, only events supported by more than half of the different methods are reported. Nucleotide numbering corresponds to the aligned sequences. Abbreviations: NRS: number of recombination sites, GIRE: genomic interval of recombination events (the span of sequences in the genome of the virus where recombination events were predicted).

Table 3. Determination of inferred putative recombination events and their frequency along the sequences of RNA 2 of the comoviruses and fabaviruses

Recombination determined by RECCO								Recombination determined by RDP v3.31β
Virus. isolate			Length of breakpoint				Genomic position of the longest breakpoint (size in nucleotide)	putative parentals (Major x Minor)
	NRS	GIRE (nt)	1 residue	2 residues	3 residues	>3residues		
BPMV.KY G-7	1	3847	1	0	0	0	3847 (1)	CPSMV x RCMV
CPMV	25	1303-3933	10	8	1	6	2201-2107 (7)	RCMV x CPSMV SqMV.CH99-211 x CPSMV
CPSMV	59	423-4154	13	17	7	22	2616-2643 (28)	RCMV x SqMV Arizona Z
SqMV.CH 99 .211	1	265-269	0	0	0	1	265-269 (1)	BPMV k-H1 x CPSMV
RCMV.S	38	521-4018	13	9	7	9	1961-1974 (14)	SqMV.CH99-211 x CPSMV CPSMV x SqMV.Y CPMV x BPMV k-H1
SqMV.Kimble	1	307	1	0	0	0	307 (1)	BPMV.kH1 x CPSMV
SqMV.Arizona Z	1	268	1	0	0	0	268 (1)	BPMV.kH1 x CPSMV
BBWV2.IA	2	22-177	0	0	0	0	22-48 (27)	/
BBVW2.MB7	1	174-194	0	0	0	0	174-194 (21)	/
BBWV2.ME	6	57-2120	2	0	1	3	79-94 (16)	/
GMV	34	60-3387	8	7	6	13	2016-2030 (15)	BBWV1.ATCC PV132 x BBWV2. MB7

With RDP v3.31β algorithm, only events supported by more than half of the different methods are reported. Nucleotide numbering corresponds to the aligned sequences. Abbreviations: NRS: number of recombination sites, GIRE: genomic interval of recombination events (the span of sequences in the genome of the virus where recombination events were predicted).

Table 4. Determination of inferred putative recombination events and their frequency along the sequences of RNA 2 of the nepoviruses

Recombination determined by RECCO								Recombination determined by RDP v3.31β
Virus. isolate			Length of breakpoint				Genomic position of the longest breakpoint (size in nucleotide)	putative parentals (Major x Minor)
	NRS	GIRE (nt)	1 residue	2 residues	3 residues	>3residues		
BRSV. S	31	1608-5994	15	6	0	10	3061-3092 (32)	GCMV x GARSV GARSV x GCMV TBRV.MJ x GCMV
CNSV	52	1996-5770	20	8	6	18	5267-5276 (10)	/
GCMV	28	1754-5861	13	6	4	5	3374-3413 (40)	TBRV.MJ x BRSV.S BRSV.S x TBRV.ED GARSV x TBRV.MJ
OLRSV	59	2041-5728	26	11	6	16	4605-4619 (15)	/
BCRV	57	1939-5776	24	9	5	19	5767-5775 (9)	/
ToRSV	60	2045-5775	19	18	6	17	4118-4133 (16)	OLRSV x GARSV
GARSV	31	1689-5861	17	3	2	9	1941-1951 (11)	GCMV x BRSV.S GCMV x TBRV.MJ TBRV.MJ x BRSV.S
GDV	37	1783-5988	10	2	5	20	3900-3926 (27)	ArMV.Butterbur x GFLV.Vol47c3 ArMV.Lv x GFLV.NW ArMV.NW x ArMV.LV ArMV.Butterbur x GFLV.A17b ArMV.Lilly x GFLV.A17d ArMV.Butterbur x GFLV.Vol50c2 ArMV.NW x GFLV.Vol47c5 ArMV x GFLV.A17d
ArMV. NW	15	1875-5024	4	0	1	10	4933-4981 (49)	ArMV.Lv x GFLV.Vol52c1 ArMV.BUtterburxGFLV.Vol54c2 GDV x GFLV.Vol55c3 ArMV.Lilly x GFLV.Vol55c3 ArMV.Butterbur x GFLV.Vol57c4 ArMV.Lilly x GFLV.Vol52c1 GFLV.Vol52c1 x ArMV.Butterbur ArMV.Lv x GFLV.Vol54c2 ArMV.Lilly x GFLV.A17b GDV x GFLV.NW GFLV.Vol51c3 x ArMV.Lv

Recombination determined by RECCO								Recombination determined by RDP v3.31β
Virus. isolate			Length of breakpoint				Genomic position of the longest breakpoint (size in nucleotide)	putative parentals (Major x Minor)
	NRS	GIRE (nt)	1 residue	2 residues	3 residues	>3residues		
ArMV. Lilly	3	2604-4015	0	0	0	3	3446-3472 (27)	ArMV.ButterburxGFLV.Vol54c2 ArMV.NW x GFLV.A17d ArMV.ButterburxGFLV.Vol57c4 ArMV.ButterburxGFLV.Vol52c1 ArMV.Lv x ArMV.NW GDV x GFLV.NW
ArMV. Narcissus	3	2565-3848	0	0	0	3	2565-2577 (13)	ArMV.ButterburxGFLV.Vol54c2 ArMV.NW x GFLV.A17d ArMV.ButterburxGFLV.Vol55c1 ArMV.ButterburxGFLV.Vol52c1 GDV x GFLV.NW ArMV.Lilly x ArMV.ArMV.Lv
ArMV. Butterbur	29	2046-5813	7	4	6	12	3030-3083 (54)	ArMV.LV x GFLV.A17b ArMV.Narcissus x GFLV.Vol52c1 GFLV.Vol54c1 x GDV
ArMV. Lv	6	1882-2951	2	0	1	3	2294-2332 (39)	ArMV.NW x GFLV.A17d GFLV.Vol52c1xArMV.Butterbur GDV x GFLV.NW
TRSV. Bud blight	76	1961-5759	33	16	10	17	2017-2037 (21)	/
RpRSV. Cherry	8	1909-5813	2	1	0	5	3444-3457 (14)	RpRSV.RAC 815 x RpRSV RpRSV.grapevine x RpRSV RpRSV x RpRSV.grapevine RpRSV x RpRSV.RAC 815
RpRSV	13	1919-6036	6	2	0	5	2160-2202 (44)	RpRSV.Cherry x RpRSV.grapevine RpRSV.Cherry x RpRSV.RAC815
RpRSV. RAC 815	6	2017-5804	2	0	2	2	2017-2081 (65)	RpRSV.grapevine x RpRSV.Cherry
RpRSV. grapevine	6	1963-5939	0	1	3	2	3801-3821 (21)	/
GFLV. F 13	21	2251-2264	3	1	1	16	5245-5265 (21)	/
GFLV. B19a	14	2226-4631	0	1	3	10	3188-3264 (77)	GFLV.A17b x GFLV.A17a GFLV.Vol57c4 x GFLV.A17d GFLV.Vol51c5 x GFLV.A17d

Table 4. (Continued)

Recombination determined by RECCO								Recombination determined by RDP v3.31β
Virus. isolate			Length of breakpoint				Genomic position of the longest breakpoint (size in nucleotide)	putative parentals (Major x Minor)
	NRS	GIRE (nt)	1 residue	2 residues	3 residues	>3residues		
GFLV. A10a	3	5536-5580	0	0	0	3	5548-5574 (27)	GFLV.Vol57c4 x GFLV.A17d GFLV.Vol57c2 x GFLV.B19a GFLV.A17d x GFLV.B19a GFLV.A17a x GFLV.A17d GFLV.B19a x GFLV.A17d GFLV.A17b x GFLV.B19a GFLV.Vol51c5 x GFLV.A17d
GFLV. A17d	14	2002-5125	3	1	1	9	2487-2520 (34)	GFLV.Vol54c1 x ArMV.Lilly GFLV.F13 x ArMV.Lilly GFLV.A17a x GFLV.B19a GFLV.Vol54c2 x GFLV.B19a
GFLV. A17b	3	2576-4649	0	0	0	3	4627-4649 (23)	GFLV.Vol57c4 x GFLV.A17d GFLV.B19a x GFLV.A17a GFLV.B19a x GFLV.A10a GFLV.A10a x GFLV.B19a
GFLV. A17a	1	5371-5409	0	0	0	1	5371-5409 (39)	GFLV.A10a x A17d GFLV.Vol57c2 x GFLV.B19a GFLV.A17d x GFLV.B19a GFLV.A17b x GFLV.A17d GFLV.A17b x GFLV.B19a GFLV.A17d x GFLV.A10a GFLV.Vol51c5 x GFLV.A17d
GFLV. NW	22	1912-6003	7	1	3	11	4139-4199 (61)	/

With RDP v3.31β algorithm, only events supported by more than half of the different methods are reported. Nucleotide numbering corresponds to the aligned sequences. Abbreviations: NRS: number of recombination sites, GIRE: genomic interval of recombination events (the span of sequences in the genome of the virus where recombination events were predicted).

RDP package did not confirm these results. Within the *Comovirus* genus, the most frequently recombining virus was CPSMV (97 putative recombination sites), whereas the opposite was the virus RCMV.S (78 sites). Similarly, four out of nine fabaviruses were potential recombinants (i.e.BBWV1.ATCC PV132, BBWV2.ME, BBWV2.B935, GMV.N-1).

Conversely, BBWV1.Ben, BBWV1.Singapore, BBWV2.IA, BBWV2.MB7 and GMV.N-1 were not recombinants. Similarly, as mentioned above, RDP v3.31β algorithm did not allow the detection of any recombinant signal. Among the *Fabavirus* genus members, the most frequently recombining virus was GMV.N-1 (74 sites), while BBWV1.ATCC PV132 was the lesser frequently recombining virus (two sites). Recombination investigations of the genus *Nepovirus* based on RECCO analysis showed that all of the 13 members known so far were recombinants (i.e. ArMV.NW, ArMV.Lv, TRSV.Bud Blight, RpRSV.Cherry, RpRSV.grapevine, GFLV.F 13, BRSV.S, CNSV, GCMV, TBRV.MJ, BCRV, PRMV.Michigan grapevine, ToRSV) (Table 2). These results were congruent with RDP package output except for ArMV.Lv, RpRSV.Cherry, RpRSV.grapevine, BCRV, PRMV.Michigan grapevine and ToRSV. Although the most frequently recombining virus was BCRV (215 sites), the lesser was RpRSV (three sites). It should be noted that most of the methods incorporated in RDP package determined that TRSV and BPMV, were the essential possible major donators of the genetic information to 60% of the investigated accessions, as the major and the minor parentals, respectively. Seeking for the recombination frequency in RNA 1 of the *Comovirinae*, two-thirds of the recombinant comoviruses (i.e. CPSMV, RCMV.S) showed that in most cases their breakpoint length was constituted by a single residue. By contrast, the breakpoint interval of the greater number of putative recombination sites of CPMV was between 3 and 16 nucleotides (Table 2). In 93% of investigated nepoviruses, the breakpoint length of the greater number of putative recombination sites was limited to a single residue. In majority, the breakpoint interval of the fabaviruses was between three and 42 residues (BBWV2.ME).

Concerning RNA 2 of the *Comovirinae*, only one half of the studied comoviruses were recombinants according to RECCO. RDP package confirmed all these results and demonstrated that CPSMV (the highest recombinant virus with 59 possible recombination sites) was the major donator of the heredity material (Table 3). Similarly, with RECCO, the same recombination rate was observed in fabaviruses but RDP package confirmed these results only in the case of GMV which showed a high degree of recombination with 34 possible recombination sites. Recombination events occurred also in RNA 2 of 25 out of 55 analyzed nepoviruses. In determining potential recombination events, a high consistency was observed between RECCO and RDP package. Nonetheless, no confirmation was given by RDP program for the following viruses: CNSV, OLRSV, BCRV, RpRSV.grapevine, GFLV.F 13 and GFLV.NW. On the recombination frequency, approximately, one half of members of *Comovirus* and *Fabavirus* genera showed breakpoints length consisting of a single residue (Table 3). Conversely, about 60% of the analyzed nepoviruses displayed recombination sites having sizes overstepping three residues. The highest size was reached by GFLV.B19a isolate (77 nt) (Table 4).

Nucleotide Sequence Analysis

Maximum composite likelihood estimate of the pattern of nucleotide substitution were conducted in MEGA4.1β. The results for the *Comovirinae* were as follows: (i) RNA 1: rates of different transitional substitutions varied from 10.72 to 29.84, and those of transversional substitutions varied from 2.71 to 4.41. The nucleotide frequencies were as follows: 0.269 (A), 0.301 (T/U), 0.185 (C), and 0.245 (G). The transition/transversion rate ratios were $k_1 = 2.975$ (purines) and $k_2 = 6.759$ (pyrimidines). The overall transition /transversion bias was R = 2.043, where $R = [AGk_1 + TCk_2]/[(A+G)(T+C)]$. There were a total of 5087 positions in the final dataset. (ii) RNA 2: rates of different transitional substitutions varied from 1.70 to 33.30, and those of transversional substitutions varied from 3.11 to 4.15. The nucleotide frequencies were as follows: 0.254 (A), 0.281 (T/U), 0.211 (C), and 0.254 (G). The transition/transversion rate ratios were $k_1 = 8.879$ (purines) and $k_2 = 0.548$ (pyrimidines). The overall transition/transversion bias was $R = 2.435$. There were a total of 2419 positions in the final dataset. In all these analyses, codon positions included were first + second + third + noncoding. All positions containing gaps and missing data were eliminated from the dataset (complete deletion option).

Positive Selection

The high genetic stability of viruses can be attributed to negative or purifying selection to maintain the functional integrity of the viral genome. The degree of negative selection in genes, or the degree of functional constraint for the maintenance of the encoded protein sequence, can be estimated, as indicated above, by the ratio between the nucleotide diversities in nonsynonymous and synonymous positions (d_N/d_S). For most coding genes the d_N/d_S ratio is below unity which is consistent with negative selection against protein change. In contrast, a d_N/d_S ratio above unity may be an indication that adaptive or positive selection is driving gene divergence. In this study, pairwise comparisons of all screened accessions showed that, none of members of the subfamily *Comovirinae* had its RNA 1 under positive selection. In contrast, all studied virus genera contained members whose RNA 2 was under positive selection. In fact, 131 accession pairs involving 66 lineages were identified (details in supplementary material). In *Comovirus* genus, except BPMV (IL-Carbondale 1/Sg.II, Hanckock 1, Hopkins 1 isolates), all remaining accessions were under positive selection representing a rate of 76% (10/13). Similarly, about 82% (45/55) members of *Nepovirus* genus were under positive selection. It is worth pointing out that the identified nepoviruses belonged to the subgroups A, B, and C together with the unclassified nepovirus GDV. They included several GFLV accessions originated from eastern Europe (Slovenia). In *Fabavirus* genus, all investigated accessions were under positive selection. It is noteworthy that RCMV.S, BBWV-1.Singapore, and SqMV.Y were the most frequent viruses found under positive selection. All these results were obtained by testing the neutrality in sequence pairs with Fisher's Exact Test. The probability of rejecting the null hypothesis of strict-neutrality ($d_N = d_S$) in favor of positive selection for each sequence pair was determined. Values of P less than 0.05 were considered significant at the 5% level. The variance of the difference (d_N-d_S) was computed using the bootstrap method (500 replicates) (details in supplementary material). All analyses were conducted using the Nei-Gojobori method incorporated in

MEGA program. All positions containing gaps and missing data were eliminated from the dataset (complete deletion option). There were a total of 268 positions in the final dataset.

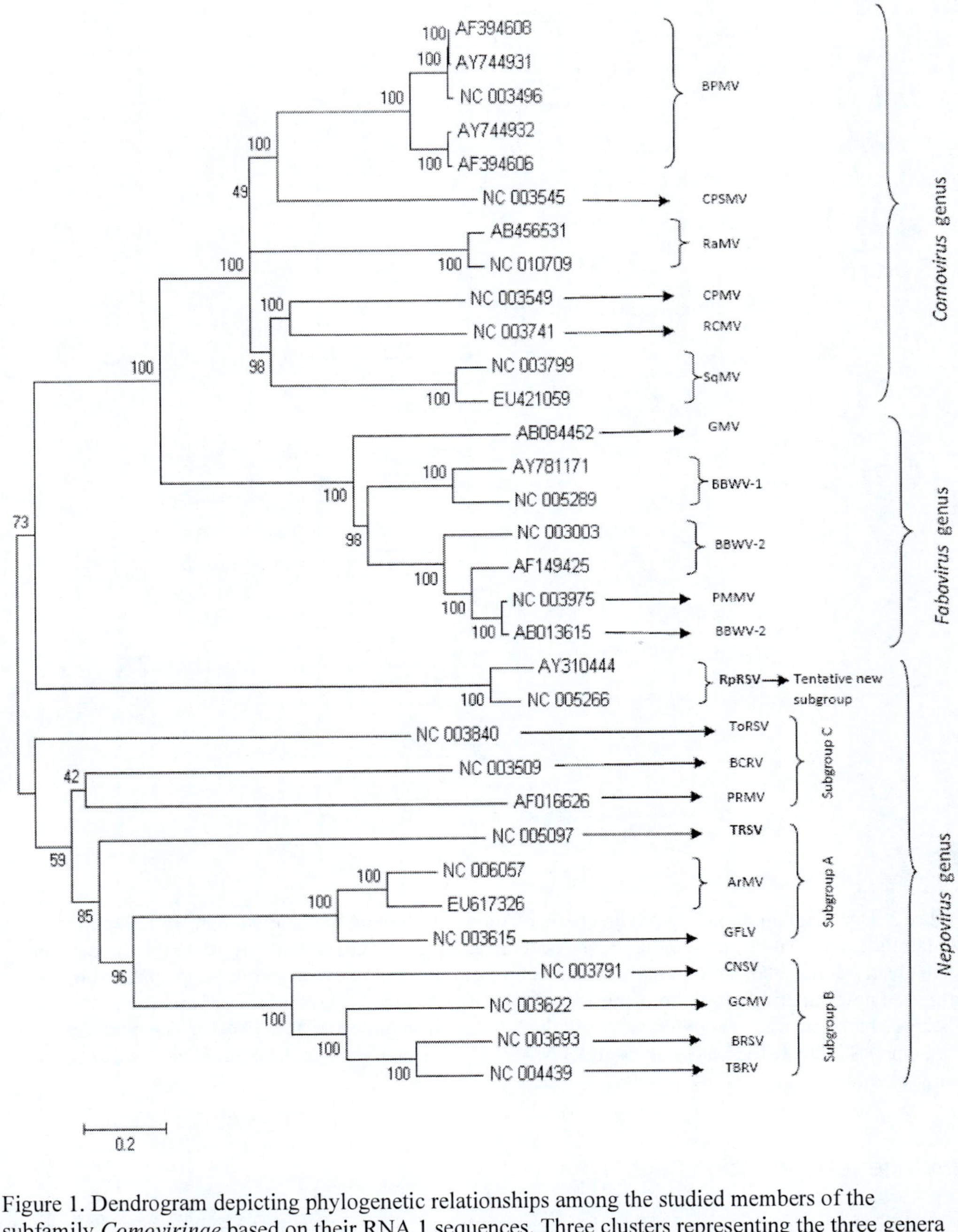

Figure 1. Dendrogram depicting phylogenetic relationships among the studied members of the subfamily *Comovirinae* based on their RNA 1 sequences. Three clusters representing the three genera were clearly delineated. Three subgroups (A,B,C) forming the *Nepovirus* genus supplemented by the proposed new tentative subgroup comprising *Raspberry ringspot virus* and its variants were distinguished. The tree was produced using the NJ algorithm option of MEGA4.1β. The numbers above the branches indicate the bootstrap confidence value. The scale bar shows the number of substitutions per nucleotide.

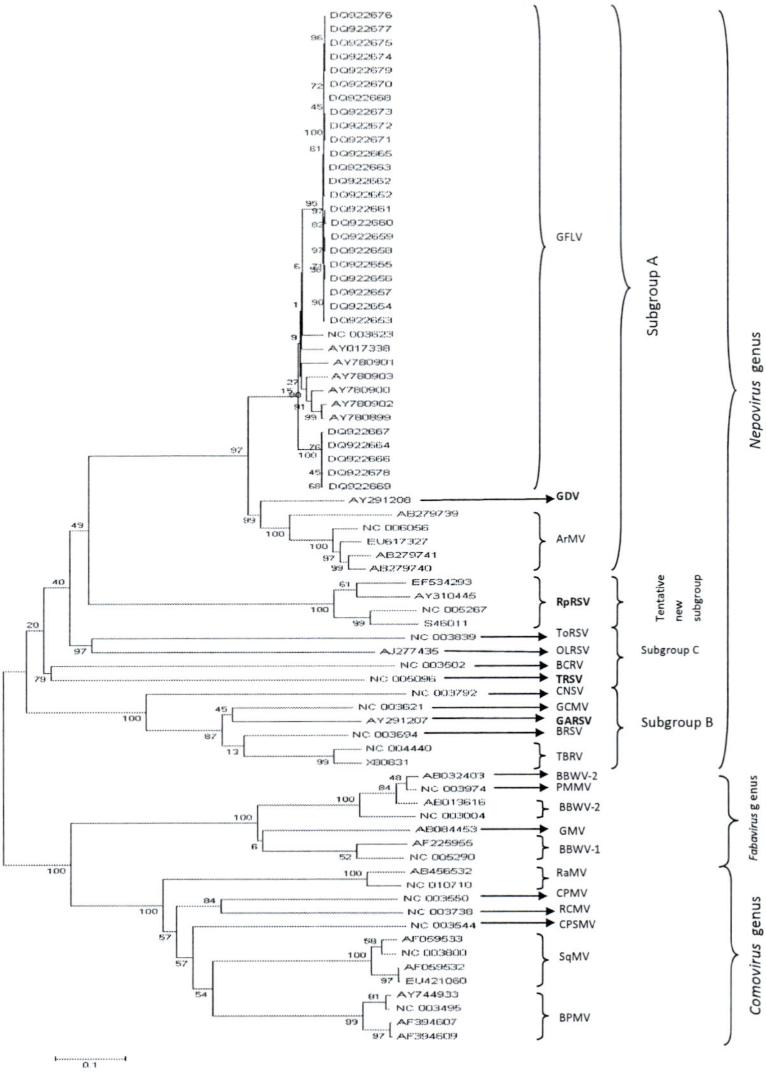

Figure 2. Dendrogram depicting phylogenetic relationships among the studied members of the subfamily *Comovirinae* based on their RNA 2 sequences. Three clusters representing the three genera were clearly delineated. Three subgroups (A,B,C) forming the *Nepovirus* genus supplemented by the proposed new tentative subgroup comprising *Raspberry ringspot virus* and its variants were distinguished. The tree was produced using the NJ algorithm option of MEGA4.1β. The numbers above the branches indicate the bootstrap confidence value. The scale bar shows the number of substitutions per nucleotide.

Phylogenetic Relationships

The phylogenetic relationships among members of the subfamily Comovirinae based on their RNAs 1 and 2 sequences were inferred using NJ, ME, MP, ML, and UPGMA. The constructed trees basically gave a somewhat similar phylogenetic grouping for the five methods used in this study. As a general picture, RNAs 1 and 2 showed phylogenies fully congruent except for TRSV. According to RNA 1, TRSV is phylogenetically closely related to ArMV and GFLV and all form the subgroup A of the *Nepovirus* genus (Figure 1). According to RNA 2, TRSV is a closer relative of ToRSV, OLRSV, and BCRV and all compose the subgroup C of the same genus (Figure 2). Interestingly and based on this disjunction, TRSV is presumably subjected to reassortment events. Remarkably, even with this discrepancy, each taxonomical genus in the subfamily *Comovirinae* constituted a homogenous group clearly distinct from one another. Based on the inferred evolutionary history of members of *Nepovirus* genus, three main clusters were delineated: subgroups A, B, and C. Moreover, a supplementary subgroup comprising various isolates of RpRSV was pointed out as a distinct cluster from the subgroups A, B, and C in the *Nepovirus* genus and to which we allocated the name: tentative new subgroup.

Discussion

Studies of the molecular evolutionary history of viruses help to provide an understanding of important features of their biology such as changes in virulence and geographical ranges and their emergence as new epidemics, information that is essential for designing strategies for controlling viruses. Nonetheless, among evolutionary driving mechanisms, recombination can mislead the phylogenetic estimation procedure (Posada and Crandall 2002) and distort subsequent inferences based on inferred phylogenies (Schierup and Hein 2000a,b). Consequently, an essential step in any phylogeny-based analysis is to screen for and quantify evidence for recombination (Kosakovsky Pond et al. 2006). In recent years, there has been an increased interest in understanding the role of recombination in the evolution of field populations of RNA plant viruses. Recombination may result in the exchange of long nucleotide sequences and it could have bigger phenotypic effects than most mutations. This could jeopardize the efficiency of current control strategies, particularly so for the use of resistance to viruses bred in crop varieties (Garcia-Arenal and McDonald 2003). In this study, we demonstrated the occurrence of putative recombination events in several members of the subfamily *Comovirinae* and determined their phylogenetic relationships. Recombination, as an important driving force in virus evolution, has shown to influence considerably the evolutionary history of viruses as we recently reported for *Prunus necrotic ringspot virus* (PNRSV) (Boulila 2009b). Moreover, in this article we demonstrated that recombination can orientate the phylogenetic reconstruction of viruses. This was particularly clear for TRSV. According to RNA 1, TRSV was a putative recombinant and was shown to be closely related to ArMV and GFLV. The RDP program output analysis indicated that ArMV is the potential major parent of this virus. In contrast, according to RNA 2 and based on RDP package analysis, TRSV was not a possible recombinant and was shown to be a closer relative of non-recombinant viruses i.e. OLRSV, and BCRV. Moreover, TRSV is suggested to be a probable reassortant based on the observed disjunction between its RNAs 1 and 2. Furthermore, ArMV and GFLV were shown to be genetically closely related by exchanging the heredity material from one another and within themselves. This connection has been exploited earlier in planning control strategies based on cross-protection. In fact, Huss et al. (1989) found that ArMV.S clearly protected *Chenopodium quinoa* from a super infection with GFLV.F 13 as shown by a decrease or even a suppression of the synthesis of the coat protein and the nucleic acids of the GFLV.F 13 isolate. In the present study, the phylogenetic analysis showed also that GDV can play the role of an acceptor as well as a donator of segments to ArMV and GFLV (Table 4). A similar situation was observed with GARSV which is revealed to be genetically closely related to GCMV, BRSV.S, and TRSV. Therefore, at least two taxonomic implications can be drawn from the results obtained: (i) RpRSV should be genetically considered as an independent virus from members of the current subgroup A of the *Nepovirus* genus used so far by the International Committee on Taxonomy of Viruses (ICTV); (ii) GDV and GARSV should be included in the subgroups A and B of the same genus, respectively, instead of being considered as two unclassified nepoviruses. As demonstrated earlier, the use of the evolutionary history of viruses is a reliable approach for classification. Serological reactivity is one among many other criteria used to classify viruses. However, as reported by Tzanetakis and Martin (2005), serological relationships are not always reliable for assigning viruses into groups since some epitopes result from secondary and tertiary protein structures.

On the other hand, the screening of the selection pressures that have shaped genetic variation form a major part of many studies of molecular evolution. A common approach to this problem involves estimating the rates of nonsynonymous (d_N) and synonymous (d_S) substitutions. Kimura (1977) was the first to point out that their comparison can be a powerful tool for studying natural selection. Estimates of d_N significantly different from d_S, provide convincing evidence for nonneutral evolution. This approach is attractive as it does not make any assumptions regarding the demographic history of the population, unlike many neutrality tests [e.g. Tajima's test (1989)] which compare estimates of effective population size obtained using different measures of genetic variation. In this study, selective pressure investigations acting on protein expression of virus genes led to the identification of positive selection only in the RNA 2 segment of most members of the subfamily *Comovirinae*. It is worth noting that numerous viruses evolved under both mechanisms: recombination and positive selection. In fact, these were: comoviruses (BPMV, CPMV, CPSMV, RCMV, and SqMV), fabaviruses (BBWV, GMV), and nepoviruses (ArMV, TRSV, RpRSV, GFLV, BRSV, CNSV, GCMV, OLRSV, BCRV, ToRSV, GARSV, GDV). A synergism might be occurring. Such a synergism between recombination and natural selection may have played a major role in Darwinian molecular evolution.

An accurate detection of recombination from DNA sequences becomes very relevant and indeed a number of methods have been developed for that purpose. Regarding RECCO, we showed in this study that it is capable to detect recombinants in all the studied genera of the *Comovirinae*. Also, it determined the frequency of recombination events along the aligned sequences. In fact, using RECCO, we demonstrated in a previous work (Boulila 2009b) that PNRSV can exchange during recombination evolution segments of the CP gene having a size of as long as 100 residues contrary to, for instance, its RNAs 1 and 2 which exchanged only segments not exceeding 18 and four residues, respectively (Boulila 2009b). Similarly, *Prune dwarf virus* (PDV) exchanged segments in the CP gene having a size of 196 nucleotides, while its RNAs 1 and 2 did not exceed 21 and 17 residues, respectively (Boulila 2009a,c; 2010). Consequently, this was a clear indication that CP genes evolved more rapidly than genes coding for nonstructural proteins. This was consistent with the statement of Zimmern (1988) and Koonin and Gorbalenya (1989). Conversely, the present study on viruses belonging to the subfamily *Comovirinae* did not demonstrate such a statement. By contrast to RECCO, RDP package did not allow to reach such conclusions. In addition, the results obtained in both studies by RDP package were in some cases in line with those of RECCO and in others not so. In the author's opinion, the main advantage of RDP 3.31β algorithm was to indicate the potential donators of the genetic information to recombinants. In a previous study (Boulila 2007), we dealt with PHYLPRO method (Weiller 1998) which demonstrated to be limited in that it was able to indicate only the position of the breakpoint represented by the value of the broken residue. Based on these observations, although RECCO is likely to give the most satisfactory results, both RECCO and RDP package should be considered as complementary methods, however.

Finally, to author's knowledge, this article reports the largest study on recombination potentially occurring in the RNAs 1 and 2 of all currently the known members of the subfamily *Comovirinae,* the positive selection operating on amino acid sites as well as their phylogenetic reconstruction. In addition, a reclassification based on their predicted evolutionary history, was proposed.

References

Aaziz R, Tepfer M (1999) Recombination in RNA viruses and in virus-resistant transgenic plants. *J. Gen. Virol.* 80: 1339-1346.

Alejska M, Kurzyniska-Kokorniak A, Broda M, Kierzek R, Figlerowicz M (2001) How RNA viruses exchange their genetic material. *Acta Bioch. Polon.* 48: 391-407.

Awadalla P (2003) The evolutionary genomics of pathogen recombination; *Nat. Rev. Genet.* 4(1): 50-60.

Baroth M, Orlich M, Thiel HJ, Becher P (2000) Insertion of cellular NEDD8 coding sequences in a pestivirus. *Virology* 278: 456-466.

Boni MF, Posada D, Feldman MW (2007) An exact nonparametric method for inferring mosaic structure in sequence triplets. *Genetics* 176: 1035-1047.

Boulila M (2007) Phylogeny and genetic recombination of *Grapevine fanleaf virus* isolates from naturally infected vineyards in Tunisia. *Phytopathol. Mediterr.* 46: 285-294.

Boulila M (2009a) Molecular characterization of an almond isolate of *Prune dwarf virus* in Tunisia: putative recombination breakpoints in the partial sequences of the coat protein-encoding gene in isolates from different geographic origin. *Phytopathol. Mediterr.* 48: 411-421.

Boulila M (2009b) Molecular evidence for recombination in *Prunus necrotic ringspot virus*; *Plant Mol. Biol. Rep.* 27: 189-198

Boulila M (2009c) Recombination structure and genetic relatedness among members of the family *Bromoviridae* based on their RNAs 1 and 2 sequence analyses. *Virus Genes* 38(3); 435-444.

Boulila M (2010) Putative recombination events and evolutionary history of five economically important viruses of fruit trees based on the coat protein-encoding gene sequence analysis *Biochem. Genet.* 48: 357-375.

Costa-Mattioli M, Cristina J, Romero H, Perez-Bercof R, Casane D, Colina R, Garcia L, Vega I, Glikman G, Romanowsky V, Castello A, Nicand E, Gassin M, Billaudel S, Ferre V (2002) Molecular Evolution of Hepatitis A virus: a new classification based on the complete VP1 protein *J. Virol.* 76: 9516-9525.

de Villiers EM, Fauquet C, Broker TR, Bernard HU, zur Hausen H (2004) Classification of papillomaviruses *Virology* 324: 17-27.

Domingo E, Holland J, Biebricher C, Eigen M (1995) Quasispecies: the concept and the world. In: Gibbs, A.J., Calisher, C.H., Garcia-Arenal, F., (eds), Molecular Basis of virus evolution. Cambridge University Press, pp. 181-191.

Eigen M (1996) On the nature of virus quasispecies. *Trends Microbiol.* 4: 216-218.

Fitch WN (1971) Toward defining the course of evolution minimum change for a specific tree topology. *Syst. Zool.* 20: 406-416.

Fuchs M, Pinck M, Serghini MA, Ravelonandro M, Walter B, Pinck L (1989) The nucleotide sequence of satellite RNA in grapevine fanleaf virus, strain F 13. *J. Gen. Virol.* 70: 955-962.

Galtier N, Gouy M, Gautier C (1996) SEAVIEW and PHYLO_WIN: two graphic tools for sequence alignment and molecular phylogeny; *CABIOS* 12(6): 543-548.

Garcia-Arenaln F, McDonald BA (2003) An analysis of the durability of the resistance to plant viruses. *Phytopathology* 93: 941-952.

Gibbs MJ, Armstrong JS, Gibbs AJ ((2000) Sister-scanning: a Monte Carlo procedure for assessing signals in recombinant sequences *Bioinformatics* 16: 573-582.

Greene AE, Allison RF (1994) Recombination between viral RNA and transgenic plants transcripts. *Science* 263: 1423-1425.

Hartigan JA (1973) Minimum mutation fits to a given tree *Biometrics* 29: 53-63.

Holland JJ, DeLaTorre JC, Steinhauer DA (1992) RNA virus populations as quasispecies. In: Holland J.J. (eds.), Genetic diversity of RNA viruses. Berlin-Springer-Verlag, pp. 1-20.

Huss B, Walter B, Fuchs M (1989) Cross-protection between *arabis mosaic virus* and *grapevine fanleaf virus* isolates in *Chenopodium quinoa*; *Ann. Appl. Biol.* 114(1): 45-60.

Katayama K, Shirato-Horikochi H, Kojima S, kageyama T, Oka T, Hoshino FB, Fukushi S, Shinohara M, Uchida K, Suzuki Y, Gojobori T, Takeda N (2002) Phylogenetic analysis of the complete genome of 18 Norwalk-like viruses. *Virology* 299: 225-239.

Kimura M (1977) Preponderance of synonymous changes as evidence for the neutral theory of molecular evolution. *Nature* 38: 515-524.

Koonin EV, Gorbalenya AE (1989) Evolution of RNA genomes does the high mutation rate necessitate high rate of evolution of virus protein? *J. Mol. Evol.* 28: 524-527.

Kosakovsky Pond SL, Posada D, Gravenor MB, Woelk CH, Frost SDW (2006) Automated phylogenetic detection of recombination using a genetic algorithm. *Mol. Biol. Evol.* 23(10): 1891-1901.

Kumar S, Nei M, Dudley J, Tamura K (2008) MEGA: A biologist-centric software for evolutionary analysis of DNA and protein sequences. *Brief Bioinfor.* 9(4): 299-306.

Lai MMC (1992) RNA recombination in animal and plant viruses. Microbiol Rev 56: 61-79.

Larkin MA, Blackshileds G, Brown NP, Chenna R, McGettigan PA, McWilliam H, Valentin F, Wallace IM, Wilm A, Lopez R, Thompson JD, Gibson TJ, Higgins DG (2007) Clustal W and Clustal X version 2.0. *Bioinformatics* 23(21): 2947-2948.

Legg JP, Thresh JM (2000) Cassava mosaic virus disease in East Africa: a dynamic disease in a changing environment. *Virus Res.* 71: 135-149.

Lukashov VV, Goudsmit J (2002) Evolutionary relationships among *Astriviridae*. *J. Gen. Virol.* 83: 1397-1405.

Martin D, Rybicki E (2000) RDP: detection of recombination amongst aligned sequences; *Bioinformatics* 16: 562-563.

Martin DP, Posada D, Crandall KA, Williamson C (2005a) A modified bootscan algorithm for automated identification of recombination sequences and recombination breakpoints. *AIDS Res. Hum. Retrovir.* 21: 98-102.

Martin DP, Williamson C, Posada D (2005b) RDP2: recombination detection and analysis from sequence alignments. *Bioinformatics* 21: 260-262.

Massey SE, Churbanov A, Rastogi S, Liberles DA (2008) Characterizing positive and negative selection and their phylogenetic effects. *Gene* 418: 22-26.

Maydt J, Lengauer T (2006) Recco: recombination analysis using cost optimization; *Bioinformatics* 22(9): 1064-1071.

Monci F, Sanchez-Campos S, Navas-Castillo J, Moriones E (2002) A natural recombinant between the geminiviruses tomato yellow leaf curl Sardinia virus and tomato yellow leaf curl virus exhibits a novel pathogenic phenotype and is becoming prevalent in Spanish populations. *Virology* 303: 317-326.

Nagai M, Sakoda Y, Mori M, Hayashi M, Kida H, Akashi H (2003) Insertion of a cellular sequence and RNA recombination in the structural protein coding region of cytopathogenic bovine viral diarrhea virus *J. Gen. Virol.* 84Pt2: 447-452.

Nei M, Gojobori T (1986) Simple methods for estimating the numbers of synonymous and nonsynonymous nucleotide substitutions. *Mol. Biol. Evol.* 3(5): 418-426.

Padidam M, Sawyer S, Fauquet CM (1999) Possible emergence of new geminiviruses by frequent recombination. *Virology* 265: 218-225.

Pinck L, Fuchs M, Pinck M, Ravelonandro M, Walter B (1988) A satellite RNA in grapevine fanleaf virus strain F 13. *J. Gen. Virol.* 69: 233-239.

Posadan D, Crandall K (2001) Evaluation of methods for detecting recombination from DNA sequences. Computer simulation. *Proc. Natl. Acad. Sci. USA* 98: 13757-13762.

Posada D, Crandall KA (2002) The effect of recombination on the accuracy of phylogeny estimation. *J. Mol. Evol.* 54: 396-402.

Qi J, Luo H, Hao B (2004a) CVTree: a phylogenetic tree reconstruction tool based on whole genomes. *Nucleic Acids Res.* 32: W45-W47.

Qi J, Wang B, Hao B (2004b) Whole genome prokaryote phylogeny without sequence alignment: a K-string composition approach. *J. Mol. Evol.* 58: 1-11.

Sanfaçon H, Wellink J, Le Gall O, Karasev A, van der Vlugt R, Wetzel T (2009) *Secoviridae*: a proposed family of plant viruses within the order Picornavirales that combines the family *Sequiviridae* and *Comoviridae*, the unassigned genera *Cheravirus* and *Sadwavirus*, and the proposal genus *Torradovirus*. *Arch. Virol.* 154(5): 899-907.

Schierup M, Hein J (2000a) Consequences of recombination on traditional phylogenetic analyses. *Genetics* 156: 879-891.

Schierup M, Hein J (2000b) Recombination and the molecular clock. *Mol. Biol. Evol.* 17: 1578-1579.

Smith JM (1992) Analyzing the mosaic structure of genes *J. Mol. Evol.* (34): 126-129.

Stuart GW, Moffet K, Baker S (2002a) Integrated gene and species phylogenies from unaligned whole genome protein sequences. *Bioinformatics* 18: 100-108.

Stuart GW, Moffet K, Leader JJ (2002b) A comprehensive vertebrate phylogeny using vector representations of proteins sequences from whole genome. *Mol. Biol. Evol.* 19: 554-562.

Stuart GW, Berry MW (2003) A comprehensive whole genome bacterial phylogeny using correlated peptide motifs defined in a high dimensional vector space. *J. Bioinfor. Comput. Biol.* 1: 475-493.

Suzuki Y (2004) New methods for detecting positive selection at single amino acid sites. *J. Mol. Evol.* 59: 11-19.

Suzuki Y (2006) Statistical properties of the methods for detecting positively selected amino acid sites. *Gene* 365: 125-129.

Suzuki Y, Gojobori T (1999) A method for detecting positive selection at single amino acid sites. *Mol. Biol. Evol.* 16(10): 1315-1328.

Tajima F (1989) Statistical-method for testing the neutral mutation hypothesis by DNA polymorphism. *Genetics* 123: 585-595.

Thiéry R, Cozien J, de Boisséson C, Kerbart-Bosher S, Névarez L (2004) Genomic classification of new betanodavirus isolates by phylogenetic analysis of the coat protein suggests a low host-fish species specificity. *J. Gen. Virol.* 85: 3079-3087.

Tzanetakis IE, Martin RR (2005) New features in the genus *Ilarvirus* revealed by the nucleotide sequence of *Fragaria chiloensis latent virus*. *Virus Res.* 112: 32-37.

Weiller GF (1998) Phylogenetic profiles: a graphical method for detecting genetic recombination in homologous sequences. *Mol. Biol. Evol.* 15: 326-335.

Worobey M, Holmes EC (1999) Evolutionary aspects of recombination in RNA viruses. *J. Gen. Virol.* 80: 2535-2543.

Zimmern D (1988) Evolution of RNA viruses. In: Holland, J.J., Domingo, E.R., Ahlquist, P. (eds). RNA genetics, CRC Press, Boca Raton. pp. 211-240.

Supplementary Material

Accession pairs	Virus.isolate pairs	*P* value at 5% level	d_N-d_S
NC_005267/NC_003004	RpRSV.Cherry/BBWV -2.ME	0.043	1.803
X80831/NC_003004	TBRV.ED/BBWV-2.ME	0.047	1.728
NC_003694/NC_003738	BRSV.S/RCMV.S	0.035	1.745
NC_003621/NC_003738	GCMV/ RCMV.S	0.034	1.877
X80831/NC_003738	TBRV.ED/ RCMV.S	0.049	1.824
NC_004440/NC_010710	TBRV.MJ/RaMV.Japanese	0.033	1.846
AB456532/AF225955	RaMV.California/BBWV-1.Singapore	0.046	1.914
NC_010710/AF225955	RaMV.Japanese/ BBWV-1.Singapore	0.019	2.455
AF059533/AB032403	SqMV.Kimble/BBWV-2.IA	0.033	1.876
NC_004440/AB032403	TBRV.MJ/ BBWV-2.IA	0.032	1.914
EF534293/AB032403	RpRSV.RAC815/ BBWV-2.IA	0.002	2.935
NC_004440/NC_003974	TBRV.MJ/PMMV.Philippines	0.032	1.912
EF534293/NC_003974	RpRSV.RAC815/ PMMV.Philippines	0.010	2.471
NC_005267/NC_003974	RpRSV.Cherry/ PMMV.Philippines	0.044	1.646
X80831/NC_003974	TBRV.ED/ PMMV.Philippines	0.030	1.808
X80831/AB013616	TBRV.ED/BBWV-2.MB7	0.020	2.037
X80831/NC_005290	TBRV.ED/BBWV-1.ATCC PV 132	0.018	2.184
AF059533/NC_003502	SqMV.Kimble/BCRV	0.007	2.714
NC_003550/NC_003502	CPMV/ BCRV	0.011	1.869
AF059532/NC_003502	SqMV.Arizona.Z/BCRV	0.004	2.838
AF225955/NC_003502	BBWV-1.Singapore/ BCRV	0.018	2.147
NC_005290/NC_003502	BBWV-1.ATCC PV 132/ BCRV	0.038	1.923
NC_005096/NC_003839	TRSV.Bud blight/ToRSV.Raspberry	0.030	1.821

Accession pairs	Virus.isolate pairs	P value at 5% level	dN-dS
NC_004440/AB084453	TBRV.MJ/GMV.N-1	0.002	2.993
NC_003694/AB084453	BRSV.S/ GMV.N-1	0.018	1.994
X80831/AB084453	TBRV.ED/ GMV.N-1	0.013	2.260
NC_003544/AB084453	CPSMV/ GMV.N-1	0.011	2.362
NC_003502/EU421060	BCRV/SQMV. CH 99-211	0.004	2.955
AB032403/NC_003800	BBWV-2.IA/SqMV.Y	0.037	1.781
AB013616/NC_003800	BBWV-2.MB7/SqMV.Y	0.037	1.697
NC_003502/NC_003800	BCRV/SqMV.Y	0.006	2.599
NC_003974/NC_003495	PMMV.Phillipines/BPMV.KY G-7	0.047	1.792
AB084453/NC_003495	GMV.N-1/ BPMV.KY G-7	0.037	1.944
NC_003004/AY780901	BBWV-2.ME/GFLV.A17d	0.035	2.068
AF059532/DQ922664	SqMV.Arizona.Z/GFLV.vol51c3	0.044	1.798
NC_003738/DQ922664	RCMV.S/ GFLV.vol51c3	0.037	1.666
AF225955/DQ922664	BBWV-1.Singapore/ GFLV.vol51c3	0.038	1.860
NC_003800/DQ922664	SqMV.Y/ GFLV.vol51c3	0.028	2.049
AF059533/DQ922678	SqMV.Kimble/ GFLV.vol57c5	0.049	1.795
NC_003738/DQ922678	RCMV.S/ GFLV.vol57c5	0.037	1.698
AF225955/DQ922678	BBWV-1.Singapore/ GFLV.vol57c5	0.032	1.888
NC_003800/DQ922678	SqMV.Y/ GFLV.vol57c5	0.031	2.019
AF059532/DQ922667	SqMV.Arizona.Z/GFLV.vol52c1	0.049	1.835
NC_003738/DQ922667	RCMV.S/GFLV.vol52c1	0.035	1.665
NC_003544/DQ922667	CPSMV/ GFLV.vol52c1	0.045	1.501
AF225955/DQ922667	BBWV-1.Singapore/ GFLV.vol52c1	0.029	1.944
NC_003800/DQ922667	SqMV.Y/ GFLV.vol52c1	0.024	2.058
NC_003738/DQ922666	RCMV.S/GFLV.vol51c4	0.037	1.668
AF225955/DQ922666	BBWV-1.Singapore/ GFLV.vol51c4	0.041	1.787
NC_003800/DQ922666	SqMV.Y/ GFLV.vol51c4	0.028	2.017
S46011/DQ922666	RpRSV/ GFLV.vol51c4	0.047	1.943
AF059533/DQ922669	SqMV.Kimble/ GFLV.vol54c2	0.049	1.642
NC_003738/DQ922669	RCMV.S/ GFLV.vol54c2	0.045	1.598
AF225955/DQ922669	BBWV-1.Singapore/ GFLV.vol54c2	0.032	1.883
NC_003800/DQ922669	SqMV.Y/ GFLV.vol54c2	0.046	1.852
NC_003800/AB279741	SqMV.Y/ArMV.Lilly	0.021	1.993
AF059533/AB279740	SqMV.Kimble/ArMV.Narcissus	0.011	2.230
AF059532/AB279740	SqMV.Arizona.Z/ ArMV.Narcissus	0.038	1.854
NC_003004/AB279740	BBWV-2.ME/ ArMV.Narcissus	0.043	1.854
AF225955/AB279740	BBWV-1.Singapore/ ArMV.Narcissus	0.013	2.239
EU421060/AB279740	SqMV.CH 99.211/ ArMV.Narcissus	0.035	1.873
NC_003800/AB279740	SqMV.Y/ ArMV.Narcissus	0.005	2.753
NC_003495/AB279740	BPMV.KY G-7/ ArMV.Narcissus	0.034	1.735
NC_003004/EU617327	BBWV-2.ME/ArMV.Lv	0.011	2.549
AF225955/EU617327	BBWV-1.Singapore/ ArMV.Lv	0.003	2.811
AB032403/EU617327	BBWV-2.IA/ ArMV.Lv	0.034	1.869
NC_003974/EU617327	PMMV.Philippines/ ArMV.Lv	0.013	2.287
AB013616/EU617327	BBWV-2.MB7/ ArMV.Lv	0.021	2.036
NC_005290/EU617327	BBWV-1.ATCC PV 132/ ArMV.Lv	0.035	1.882
AB084453/EU617327	GMV.N-1/ ArMV.Lv	0.039	1.875
AB279741/EU617327	ArMV.Lilly/ ArMV.Lv	0.033	2.206
NC_003004/NC_006056	BBWV-2.ME/ArMV.NW	0.038	2.016

(Continued)

Accession pairs	Virus.isolate pairs	P value at 5% level	dN-dS
AB084453/NC_006056	GMV.N-1/ ArMV.NW	0.024	1.972
AF225955/DQ922654	BBWV-1.Singapore/GFLV.vol47c2	0.045	1.749
AF225955/DQ922657	BBWV-1.Singapore/GFLV.vol47c5	0.049	1.661
AF059533/AY780903	SqMV.Kimble/GFLV.B19a	0.029	2.109
NC_003800/AY780903	SqMV.Y/ GFLV.B19a	0.011	2.461
AF225955/DQ922663	BBWV-1.Singapore/GFLV.vol51c2	0.017	2.135
NC_003800/DQ922663	SqMV.Y/ GFLV.vol51c2	0.050	1.768
NC_003738/DQ922675	RCMV.S/GFLV.vol57c2	0.047	1.483
AF225955/DQ922675	BBWV-1.Singapore/ GFLV.vol57c2	0.025	2.048
NC_003800/DQ922675	SqMV.Y/ GFLV.vol57c2	0.038	1.891
NC_003738/DQ922662	RCMV.S/GFLV.vol51c1	0.047	1.483
AF225955/DQ922662	BBWV-1.Singapore/ GFLV.vol51c1	0.025	2.048
NC_003800/DQ922662	SqMV.Y/ GFLV.vol51c1	0.038	1.891
NC_003738/DQ922665	RCMV.S/GFLV.vol51c4	0.038	1.483
AF225955/DQ922665	BBWV-1.Singapore/ GFLV.vol51c4	0.025	2.048
NC_003800/DQ922665	SqMV.Y/ GFLV.vol51c4	0.047	1.891
AF225955/DQ922670	BBWV-1.Singapore/GFLV.vol54c3	0.023	2.156
NC_003738/DQ922672	RCMV.S/GFLV.vol55c2	0.038	1.718
AF225955/DQ922672	BBWV-1.Singapore/ GFLV.vol55c2	0.017	2.134
NC_003800/DQ922672	SqMV.Y/ GFLV.vol55c2	0.027	1.908
NC_003738/DQ922673	RCMV.S /GFLV.vol55c3	0.027	1.718
AF225955/DQ922673	BBWV-1.Singapore/ GFLV.vol55c3	0.017	2.134
NC_003800/DQ922673	SqMV.Y/ GFLV.vol55c3	0.038	1.908
NC_003738/DQ922671	RCMV.S/GFLV.vol55c1	0.027	1.718
AF225955/DQ922671	BBWV-1.Singapore/ GFLV.vol55c1	0.017	2.134
NC_003800/DQ922671	SqMV.Y/ GFLV.vol55c1	0.038	1.908
NC_003738/DQ922674	RCMV.S/GFLV.vol57c1	0.047	1.483
AF225955/DQ922674	BBWV-1.Singapore/ GFLV.vol57c1	0.025	2.048
NC_003800/DQ922674	SqMV.Y/ GFLV.vol57c1	0.038	1.891
AF225955/DQ922668	BBWV-1.Singapore/GFLV.vol54c1	0.023	2.122
NC_003800/ DQ922668	SqMV.Y/ GFLV.vol54c1	0.047	1.865
NC_003738/DQ922676	RCMV.S/GFLV.vol57c3	0.047	1.521
AF225955/ DQ922676	BBWV-1.Singapore/ GFLV.vol57c3	0.025	2.084
NC_003800/ DQ922676	SqMV.Y/ GFLV.vol57c3	0.038	1.929
AF225955/DQ922679	BBWV-1.Singapore/GFLV.vol57c6	0.031	1.886
NC_005096/DQ922679	TRSV.Bud blight/ GFLV.vol57c6	0.049	1.654
NC_003738/DQ922677	RCMV.S/GFLV.vol57c4	0.047	1.483
AF225955/DQ922677	BBWV-1.Singapore/ GFLV.vol57c4	0.025	2.048
NC_003800/DQ922677	SqMV.Y/ GFLV.vol57c4	0.038	1.891
AF225955/DQ922652	BBWV-1.Singapore/GFLV.vol45c1	0.021	2.152
NC_003800/DQ922652	SqMV.Y/ GFLV.vol45c1	0.042	1.884
AF225955/DQ922661	BBWV-1.Singapore /GFLV.vol50c2	0.034	1.876
NC_005096/DQ922661	TRSV.Bud blight/ GFLV.vol50c2	0.046	1.729
NC_003800/DQ922661	SqMV.Y/ GFLV.vol50c2	0.045	1.784
AF225955/DQ922660	BBWV-1.Singapore/GFLV.vol50c1	0.028	1.886
NC_003738/DQ922659	RCMV.S/GFLV.vol49c2	0.043	1.556
AF225955/DQ922659	BBWV-1.Singapore/ GFLV.vol49c2	0.041	1.689

Accession pairs	Virus.isolate pairs	P value at 5% level	dN-dS
NC_005096/DQ922659	TRSV.Bud blight/ GFLV.vol49c2	0.038	1.819
NC_003800/DQ922659	SqMV.Y/ GFLV.vol49c2	0.032	1.936
AF225955/DQ922658	BBWV-1.Singapore/GFLV.vol49c1	0.031	1.917
NC_003800/DQ922658	SqMV.Y/ GFLV.vol49c1	0.050	1.735
AF225955/DQ922656	BBWV-1.Singapore/GFLV.vol47c4	0.034	1.876
NC_003800/DQ922656	SqMV.Y/ GFLV.vol47c4	0.045	1.784
AF225955/DQ922653	BBWV-1.Singapore/GFLV.vol47c1	0.037	1.845
NC_003800/DQ922653	SqMV.Y/ GFLV.vol47c1	0.037	1.876
NC_003738/NC_003623	RCMV.S/GFLV.F 13	0.018	1.894
NC_003738/AY780900	RCMV.S/GFLV.A17b	0.021	1.857
NC_003800/AY780900	SqMV.Y/ GFLV.A17b	0.036	1.912
NC_003800/AY291208	SqMV.Y/GDV	0.038	1.782

Accession pairs of the RNA 2 of the *Comovirinae* under positive selection with determined probability at the 5% level based on the Fisher's Exact Test of neutrality and calculated d_N-d_S.

In: Bioinformatics Research: New Developments
Editors: C. Battik and K. Belhassine
ISBN: 978-1-61942-363-3

Chapter III

Discovering Gene-Gene and Gene-Environment Causal Interactions from Different Types of Genomic Data Using Statistical Methods in Bioinformatics

Changwon Yoo
Florida International University,
Department of Epidemiology and Biostatistics,
Miami, Florida, US

Abstract

In this chapter we will review bioinformatics approaches in modeling the gene-gene and gene-environment causal interactions of systems biology studies, such as Genome-wide association studies (GWAS) and Microarray gene expression studies. Many gene-gene causal interactions models have been introduced recently.

However, given many different types of genomic data – such as gene sequence, gene expression, etc., not many bioinformatics approaches are available in combining knowledge extracted from the data. To this end, we will review gene-gene and gene-environment interaction in terms of Single Nucleotide Polymorphism (SNP) and gene expression studies.

We introduce a promising probabilistic causal model, i.e., causal Bayesian networks, and discuss how to represent gene-environment causal interactions using the probabilistic causal model. We conclude this chapter with promising research directions of gene-gene and gene-environment causal interactions models using bioinformatics approaches.

Introduction

In biomedical science, we want to find out gene-gene and gene-environment causal interactions. Ultimately, we want to find out how these causal interactions influence selected traits at any given point in life. For bioinformaticians, biostatisticians, and epidemiologists the gene-gene and gene-environment causal interactions are defined by statistical probabilities. On the other hand, for the biological scientist, the mere statistical gene-gene and gene-environment interactions would not be considered as sufficient basis for the actual biological interactions.

The primary objective of this chapter is to examine the gene-gene and gene-environment interactions from different types of genomic data using statistical and bioinformatics approaches. The gene-gene causal interactions have been modeled using high throughput data from Single Nucleotide Polymorphism (SNP) studies (Schwender and Ickstadt 2007; Chen, Sun et al. 2008; He, Oetting et al. 2009`; Gui, Andrew et al. 2011) and gene expression microarray studies (Friedman, Linial et al. 2000; Yoo and Cooper 2004; Lin, Lin et al. 2009; Pedraza, Gomez-Capilla et al. 2010; Yoo, Blitz et al. 2011). Recent research in biology shows that the gene-gene interactions cannot be described without mentioning the environment in which the interactions are taking place. Moreover, recent studies in the field of epigenetics provide us with possible gene-environment interactions can biochemically alter the genome. The complexity of a statistical model for gene-gene causal interactions is already overwhelming; however, we need to be aware that additional to the gene-gene causal interactions, gene-environment interactions should be modeled to complete the picture.

In the following sections, we will look more into different types of genomic data, i.e., gene expression data, SNP data, and epigenetic data, and review what bioinformatics and statistical approaches have been used to analyze these data.

In conclusion, we will show what traditional statistical methods and recent biostatistics methods can be used in modeling gene-gene and gene-environment interactions. Also in the conclusion of this chapter, we will discuss about a promising bioinformatics approach, i.e., causal Bayesian networks, that provides a natural way of describing causal relationships among modeled variables.

Gene Expression Data

This section is partially adopted and summarized from Yoo (Yoo 2010). Microarray techniques opened a new era of studies in molecular biology. Before these techniques existed, there were labor-intensive methods like Northern blot analyses to measure a single gene's expression patterns in cells. Current microarray techniques can measure the expression level of about 10,000 genes at a time. A successful sequencing of the entire genome of yeast *Saccharomyces cerevisiae* in April 1996 initiated many gene expression experimental studies in yeast (Spellman, Sherlock et al. 1998; Ideker, Thorsson et al. 2001; Smith and Kruglyak 2008). These studies, as well as many other studies that are related to them, fit under a new approach in biology that is called *systems biology*. Systems biology seeks in part to model large networks of cellular function, including the causal pathways that capture how genes regulate each other. A recent review of system biology is provided by Ideker et al. (2001) and Kitano (2002).

Before describing gene-gene causal interaction (genetic pathway) models, we first place them in the context of gene clustering methods, which have been very popular the last few years. Indeed most of the early work on gene expression data analyses used clustering methods. Gene expression levels that were measured by cDNA microarray in the yeast cell-division cycle were analyzed for the first time using a cluster analysis (Spellman, Sherlock et al. 1998). A cluster analysis typically searches for groups of genes that show similar expression pattern among different experimental conditions. Other analyses followed using similar cluster analyses applied to microarray data (Michaels, Carr et al. 1998; Golub, Slonim et al. 1999; Herwig, Poustka et al. 1999).

Clinical studies also used cluster analysis on microarray data (Alizadeh, Eisen et al. 2000; Getz, Levine et al. 2000). For example, Alizadeh et al. (2000) used cluster analysis to find different types of diffuse large B-cell lymphoma (DLBCL) among DLBCL patients by comparing the clusters of similarly expressed genes and whether or not they responded to the current therapy. Recently many microarray clinical studies have been published (see Section on *Human Genome Variation: Linking Genotypes to Clinical Phenotypes* in Pacific Symposium on Biocomputing 2001). Along with classification or cluster analyses, gene pathway (network) analyses were performed on the gene expression data. Analyses to construct pathways among the genes yield more information than do cluster or classification analyses. Cluster and classification analyses do not necessarily provide causal information, which is at the heart of gene pathway discovery. On the other hand, knowledge of causal pathways can be used to produce a causal clustering of the genes.

In the following subsections, we will briefly review gene-gene causal interaction (genetic pathway or network) models. More detail review can be found in Yoo (Yoo 2010).

Boolean networks: Boolean networks were first introduced by Somogyi and Sniegoski (1996). A structure search algorithm in Boolean networks called REVEAL (REVerse Engineering ALgorithm) was introduced by Liang et al. (1998). With its simple representation, Boolean networks were easily implemented as genetic networks. However, since Boolean networks do not explicitly model uncertainty, they cannot model the stochastic behavior of a biological system. Also note that Boolean networks do not use any arrows for the arcs among gene-gene and gene-environment interactions. Thus, there is no sense of causality in the model.

Continuous models: In mathematics, using differential equations to model a biological system has a long history (Goodwin 1965; Griffith 1968; Griffith 1968). Chen et al. (1999) modeled a dynamic system of genes using differential equations. They modeled a simplified dynamic system of gene regulation (with feedback on transcription). Differential equations can model biological dynamics more specifically than Boolean networks, but the computational cost of the differential equation model is high, and often many of the parameters needed in the differential equation model are not available. Since most of the dynamics of the actual genetic pathways appear to be non-linear, a linear model seems to work on only limited dynamics of the genetic pathway.

Bayesian networks: The Bayesian network (BN) model, also known as a probabilistic causal network, has been widely used to learn predictive models from data. The BN overcomes the shortcomings of the simple Bayesian method, first explored extensively in the early 1960s, that assumed conditional independence of the evidence given an hypothesis. BNs can model causality based on expert's knowledge, data or both. It is also used in many medical related domains because of its ability to perform inference in the forward direction, backward direction, or any combination (Citro, Banks et al. 1997; Chevrolat, Golmard et al. 1998; Lucas, Bruijn et al. 2000). One practical limitation of BNs is that inference within them is NP-hard (Cooper 1987), which is a frequent limitation of powerful reasoning methodologies; in response, researchers have developed approximate inference methods to apply to complex BN inference tasks.

A causal Bayesian network (or *causal network* for short) is a Bayesian network in which each arc is interpreted as a direct causal influence between a parent node (variable) and a child node, relative to the other nodes in the network (Pearl 1988). Figure 1 illustrates the structure of a hypothetical causal Bayesian network structure containing five nodes that represent genes. The probabilities associated with this causal network structure are not shown.

The causal network structure in Figure 1 indicates, for example, that the *Gene*1 can regulate (causally influence) the expression level of the *Gene*3, which in turn can regulate the expression level of the *Gene*5. The causal Markov condition gives the conditional independence relationships specified by a causal Bayesian network:

A variable is independent of its non-descendants (i.e., non-effects) given its parents (i.e., its direct causes).

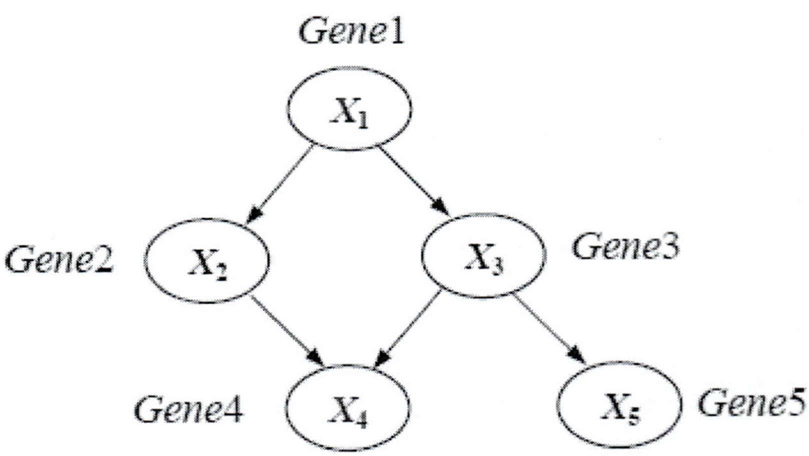

Figure 1. A causal Bayesian network that represents a hypothetical gene-regulation pathway.

The causal Markov condition permits the joint distribution of the *n* variables in a causal Bayesian network to be factored as follows (Pearl 1988):

$$P(x_1, x_2, ..., x_n \mid K) = \prod_{i=1}^{n} P(x_i \mid \pi_i, K) \tag{1}$$

where x_i denotes a state of variable X_i, π_i denotes a joint state of the parents of X_i, and K denotes background knowledge. Bayesian networks have been used in analyzing gene expression data from microarray experiments in order to model gene regulated pathways (Friedman, Linial et al. 2000; Hartemink, Gifford et al. 2001).

Murphy and Mian (1999) showed that the Boolean network model (Somogyi and Sniegoski 1996), the linear model (D'haeseleer, Wen et al. 1999), and the non-linear weighted model (Weaver, Workman et al. 1999) are all special cases of dynamic Bayesian networks (DBN). A DBN incorporates time in Bayesian networks (which is then usually called a temporal Bayesian network). Yeast cell-cycle-regulated gene expression data (Spellman, Sherlock et al. 1998) was first analyzed using Bayesian networks by Friedman et al. (2000). Hartemink et al. (2001) used Bayesian networks with annotated arcs to represent promotion and inhibition reactions. Spirtes et al. (2001) discuss issues of sample size, measurement error and averaging effects on gene expression studies using causal Bayesian networks.

Mixture models and other models: McAdams and Shapiro (1995) modeled the *E. coli* λ phage lysis-lysogeny genetic switch using a mixture of Boolean switches and continuous input-output relations. Yuh et al. (1998) modeled gene *endo16* in the sea urchin embryo with a similar hybrid model. Matsuno et al. (2000) used a Petri net that models continuous variables and analyzes the genetic switch mechanism of λ phage. Goss and Peccoud (1999) used stochastic Petri nets to model the stabilizing effect of the protein ROM on the genetic network regulating ColE1 plasmid replication.

There are many different kinds of statistical classification methods. A method commonly used for statistical classification is *k*-Nearest Neighbor (kNN), which classifies a new case by calculating the minimum distance (according to some metric) between the new case and a set of training cases. kNN has been used in areas such as the classification of radiology reports and the modeling of the immune system. Variations of kNN have more recently been used in classifying and clustering genes from large gene expression datasets (Michaels, Carr et al. 1998; Spellman, Sherlock et al. 1998; Golub, Slonim et al. 1999; Herwig, Poustka et al. 1999).

Petri nets is a formal graphical language appropriate for modeling systems with concurrency. Petri nets are based on a general theory for discrete parallel systems. Petri nets were used in guidelines for patient care flow (Quaglini, Stefanelli et al. 2001). It has also been used in modeling mechanisms in a cell (Boucher, Doisy et al. 1998; Goss and Peccoud 1999; Matsuno, Doi et al. 2000).

Genetic programming uses the three basic mechanisms that drive natural evolution — *reproduction*, *mutation*, and *selection* — in its search for a model that best fits the training data. Evolutionary methods allow a program to evolve, giving it great freedom to search through a large space of possible models. Koza et al. (2001) has used genetic programming to learn gene networks from simulated data that was generated by a computer model of the cell, called E-cell (Tomita, Hashimoto et al. 1999)

Single Nucleotide Polymorphisms

Relatively recent genome-wide association studies (GWAS) have discovered significant associations between complex diseases and certain single nucleotide polymorphisms (SNPs). A SNP is a DNA sequence variation resulting from an alteration of a single nucleotide in the genome between members of a particular species or paired chromosomes within an individual. It differs from a mutation in that the variation must occur within at least 1% of the population. SNPs are the most common genetic variations and thus are the most thoroughly investigated. It is believed that SNP-SNP interactions, not the individual SNPs themselves, play an important role in the development of complex diseases. Multiple models have been employed in SNP-SNP analysis, most notably logistic regression, combinatorial methods, support vector machines, and logic regression.

Logistic regression, a fairly traditional model used for SNP analysis, links SNPs to disease outcome using the logit function. SNP-SNP interactions can be considered by including interaction terms in the model. This of course can result in a large number of input variables, especially when there is a desire to investigate higher-order SNP interactions. If stratification is present within the data, as in the case of matched case-control studies, the conditional logistic regression (CLR) method can be used. By stratifying the matched case-control pairs, the CLR method is able to adjust for the matching of the cases and controls (Heidema, Boer et al. 2006).

A widely used combinatorial method for SNP analysis is Multifactor Dimensionality Reduction (MDR). MDR attempts to combine two or more attributes, in this case SNPs, into a single attribute to improve disease prediction. The combination of SNPs that best predicts the risk of a disease by minimizing the classification error of cases and controls is identified. A number of MDR variations have been proposed, including pair-wise MDR (PWMDR), which addresses the problem of MDR's inability to classify sparse or empty cells (He and others, 2009) and robust MDR (RMDR) which makes use of the Fisher's Exact test to determine the inclusion of SNP combinations in the overall MDR model (Gui, Andrew et al. 2011).

Goodman et al. (1999) developed an approach similar to MDR, known as polymorphism interaction analysis (PIA) to explore SNP interactions and colon cancer risk. Like MDR, PIA examines all possible SNP combinations to find the interaction that best predicts the risk of disease. It differs in that it uses two unique scoring functions to find the interactions most likely associated with disease risk, the Gini index and the percentage wrong (i.e. the percentage of misclassified subjects). In addition, PIA makes use of ten-fold cross validation and, to avoid missing data resulting in false-positive results, excludes SNPs or SNP combinations that have a case-to-control (or vice versa) ratio greater than 1.2 from the analysis (Goodman 1999).

Support vector machines (SVMs) have also been recently used in SNP-SNP analyses. SVMs are a collection of supervised learning methods used for both classification and regression. Whereas many classifiers aim to minimize prediction error, SVMs are trained to maximize prediction accuracy. Observations are represented as points in space and a hyperplane or set of hyperplanes in a high or infinite dimensional space are constructed and treated as the decision boundary between the outcome categories. The prediction accuracy is maximized by finding the hyperplane or set of hyper planes that have the greatest distance to the nearest training data points (Chen, Sun et al. 2008).

Chen et al. (2008) proposed the following four search algorithms to detect interaction among SNPs: recursive feature addition SVM (SVM-RFA), recursive feature elimination SVM (SVM-RFE), SVM with local search (SVM-Local), and SVM with genetic algorithm (SVM-GA). RFA/E discovers the optimal subset of SNP combinations by ranking the subsets according to a ranking criterion. SNP combination(s) are added/ eliminated at each iteration using the correlation coefficients as the ranking criteria (Chen, Sun et al. 2008).

The SVM-local algorithm is similar to most local searches in that a random subset of SNP interactions is initially generated. A search is then conducted through the initial subset's neighbors in an attempt to find a "better" subset. If one is found, a local move to the "better" subset is accepted. This continues until a subset is selected which has no "better" subsets in its neighborhood. For a given subset of SNP interactions, its neighbors are defined as those subsets which differ by a single element (Chen, Sun et al. 2008).

Unlike SVM-local, SVM-GA is a stochastic search that is based upon the mechanism of natural selection and genetics. The search begins by generating a random set of SNP combinations, called the population. Genetic operations, crossovers, and mutations are performed on randomly selected chromosomes (individual SNP combinations within the population) to yield the next generation. An evolution process, called the selection, is then performed on both generations to improve the chromosomes. New generations are created and the above is repeated until the chromosomes in the population converge. The final chromosome is considered the best subset of SNP interactions (Chen, Sun et al. 2008).

In logic regression, the interactions among SNPS are represented in logic trees and logic expressions. Both make use of the logic operators "or" and "and", the latter signifying an interaction. Traditional logic regression uses the Monte Carlo Markov Chain (MCMC) method to find the collection of best logic regression models. From that collection, the SNP combinations occurring most frequently are identified and assumed to be important interactions. The importance of interactions is quantified by the proportion of models in which the SNP combinations appear. Interactions that are only significant in small subgroups of the population thus have the potential to be overlooked. The LogicFS (Schwender and Ickstadt 2007) approach to logic regression uses bootstrap sampling to address this issue. Another advantage of the LogicFS approach is that it, unlike MCMC logic regression, uses two unique importance measures that allow for the comparison of interactions of different orders. Logic regression in general is considered more practical than other methods used in SNP-SNP analysis because it does not require interaction terms to be included in the model as inputs. Logistic regression, MDR, SVMs, and logic regression are all methods that are capable of identifying important SNP-SNP interactions. Algorithms that use different search mechanisms, different ranking/importance criterion, and/or that are geared toward specific situations have been proposed for each. Despite these advantages in SNP and SNP-SNP analysis, the literature seems to lack studies that seek out causal discovery among SNPs. Like the other models, Bayesian networks are able to identify important associations among SNPs. It is being proposed that Bayesian networks are also capable of extracting causal information from those SNP-SNP and SNP-disease associations.

Epigenetic Regulation of the Genome

Epigenetics modify genomes functionswithout an alteration of DNA sequence. Thus, the epigenetic modifications change the transcriptions of genes.

DNA methylation, which involves the addition of a methyl group onto cytosines in the DNA on the genome were thought to be active only during embryonic development. However, recent studies show that DNA methylation patterns are actively modified in even fully differentiated cells (Bird 2007). This shows biological examples of gene-environmental interactions. Such interactions need to be considered in modeling gene expression. The gene-environment interactions also arise from gene transcription. Transcription factor proteins bind to gene regulatory elements.

Fu et al. (2010) developed Bayesian inference methods for epigenetic data to study the transmission fidelity of DNA methylation patterns over cell divisions. Genome-wide methylation data were analyzed using the genome-wide statistical significance calculation for increased variability (Jaffe, Feinberg et al. 2011) and Bayesian hierarchical model (Wu, Yi et al. 2011). A beta-mixture model was used in analyzing genome-wide methylation patterns of colon cancer

Conclusion

We have reviewed bioinformatics and statistical methods for gene-gene and gene-environment causal interactions from different types of genomic data. Traditionally in statistics, modeling complex gene-gene and gene-environment interactions is given in the usually a linear weighted sum among variables that represent genes (G), environment (E), and gene-environment interactions (G×E) (Kraft and Hunter 2010): $f(E[Y|G,E]) = \alpha + \beta_G G + \beta_E E + \beta_{G\times E}\ G\times E$ where Y is a binary outcome, i.e., have a disease or not have the disease, E[Y|G,E] represent trait value given G and E, and f(.) represents a link function, e.g., a logit function $f(x) = \log(x) - \log(1-x)$. However, note that there are pros (e.g., explicitly modeling additive and multiplicative effect) and cons (e.g., model is not flexible) of the traditional statistical approach. Moreover, modeling causality is not a straight forward extension.

Recently, there have been many machine (or statistical) learning methods that have been applied as a bioinformatics methods in modeling complex gene-gene and gene-environment interactions. These bioinformatics methods were presented in the previous sections. Here we present causal Bayesian networks as a method that can model complex gene-gene and gene-environment interactions from different types of genomic data given the emergence of datasets in medicine and biology with large number of variables, Bayesian Networks have been successful in developing efficient algorithms that scale up to datasets involving hundreds of variables in learning high quality predictive models from genomic data (Yoo and Cooper, 2004). A Bayesian Network is a directed acyclic graph in which each node represents a variable and each arc represents probabilistic influence. In Bayesian Networks, each arc is interpreted as a direct influence between a parent node (variable) and a child node, relative to the other nodes in the network.

Bayesian networks are also build based on the causal Markov conditions (Pearl 1988), we can find out any conditional relationships among the variables in the Bayesian network (Charniak 1991). This can be understood with the following hypothetical example Bayesian network:

In the above network, either it rains (Rain) or you turn on your sprinkler (Sprinkler) can make your lawn wet (My Lawn Wet). Also if your lawn gets enough water, it gets green (My Lawn Green). Of course, your neighbor, who does not have a sprinkler, can get his lawn wet (Neighbor Lawn Wet) when it rains. In the above network, we can identify the following three sub networks:

In Figure 3 (a) (we call it *converging arcs* going into My Lawn Wet from Sprinkler and Rain), if you know your lawn is wet (My Lawn Wet) and you know it didn't rain (not Rain)

then there is a high chance that your sprinkler is on (Sprinkler), in other words, *if converging arcs into a variable C from variables A and B , then A and B becomes dependant given C.* Also note that in Figure 3 (b) (we call it *diverging arcs* from Rain into My Lawn Wet and Neighbor Lawn Wet), if it rains (Rain), your lawn (My Lawn Wet) and your neighbor lawn get wet (Neighbor Lawn Wet). If you know it rained (Rain), knowing your lawn is wet will not tell you much more about your neighbor lawn being wet, in other words, *if diverging arcs from a variable C into variables A and B, then A and B becomes independent given C.* In Figure 3 (c) (we call it *serial arcs* from Rain to My Lawn Wet to My Lawn Green), if it rains (Rain), then my lawn gets wet (My Lawn Wet), and eventually, your lawn gets green (My Lawn Green). In this case, if you know your lawn is wet (My Lawn Wet), then knowing whether it rained (Rain) or not will not tell you much more about your lawn getting green (My Lawn Green), in other words, *if serial arcs from a variable A to a variable C to a variable B, then A and B is independent given C.*

These three sub networks (converging, diverging, and serial) provide ways to express causal interactions in intuitive ways. The fact that causal Bayesian networks can provide seamless combination of prior knowledge (prior probability) with the statistical analysis of collected data (posterior probability), makes an excellent bioinformatics statistical tool in modeling complex gene-gene and gene-environment interactions from different types of genomic data.

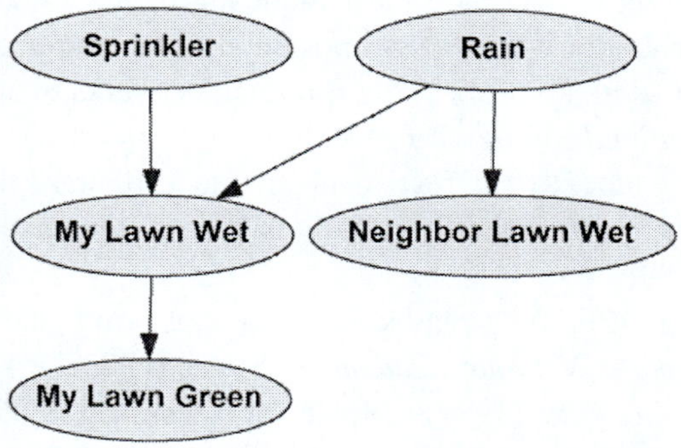

Figure 2. A simple example Bayesain network.

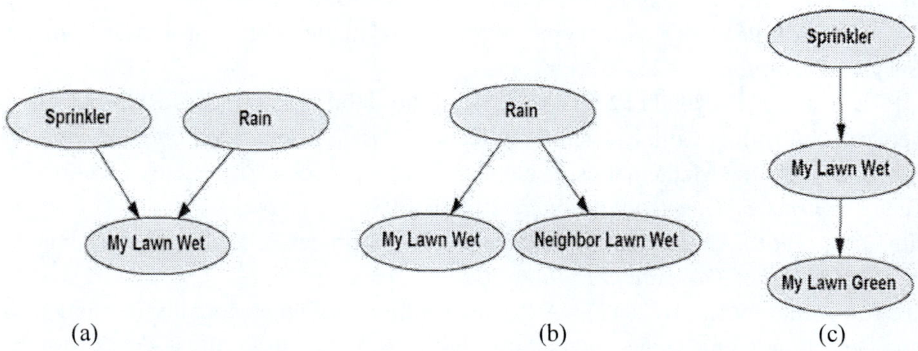

Figure 3. Sub networks from Figure 2.

References

Alizadeh, A. A., M. B. Eisen, et al. (2000). "Distinct types of diffuse large B-cell lymphoma identified by gene expression profiling." *Nature* 403(6769): 503-511.

Bird, A. (2007). "Perceptions of epigenetics." *Nature* 447: 396–398.

Boucher, A., A. Doisy, et al. (1998). "A society of goal-oriented agents for the analysis of living cells." *Artificial Intelligence in Medicine* 14(1-2): 183-199.

Charniak, E. (1991). "*Bayesian networks without tears.*" 12(4): 50–63.

Chen, S. H., J. Sun, et al. (2008). "A support vector machine approach for detecting gene-gene interaction." *Genetic Epidemiology* 32: 152-167.

Chen, T., H. L. He, et al. (1999). Modeling gene Expression with Differential Equations. *Pacific Symposium on Biocomputing.*

Chevrolat, J., J. Golmard, et al. (1998). "Modelling behavioral syndromes using Bayesian networks." *Artificial Intelligence in Medicine* 14(3): 259-277.

Citro, G., G. Banks, et al. (1997). "INKBLOT: A neurological diagnostic decision support system integrating causal and anatomical knowledge." *Artificial Intelligence in Medicine* 10: 257-267.

Cooper, G. F. (1987). Probabilistic inference using belief networks is NP-hard, Stanford University.

D'haeseleer, P., X. Wen, et al. (1999). Linear modeling of mRNA expression levels during CNS development and injury Pacific Symposium on Biocomputing, 41-52.

Friedman, N., M. Linial, et al. (2000). "Using Bayesian networks to analyze expression data." *Journal of Computational Biology* 7: 601-620.

Fu, A., D. Genereux, et al. (2010). "Statistical inference of transmission fidelity of DNA methylation patterns over somatic cell divisions in mammals." *Annals of Applied Statistics* 4(2): 871-892.

Getz, G., E. Levine, et al. (2000). "Coupled two-way clustering analysis of gene microarray data." *Proceedings of the National Academy of Sciences* 97(22): 12079-12084.

Golub, T. R., D. K. Slonim, et al. (1999). "Molecular classification of cancer: class discovery and class prediction by gene expression monitoring." *Science* 286: 531-537.

Goodman, S. N. (1999). "Probability at the bedside: The knowing of chances or the chances of knowing?" *Annals of Internal Medicine* 130: 604-606.

Goodwin, B. C. (1965). "Oscillatory behavior of enzymatic control processes." *Advances in Enzyme Regulation* 3: 425-439.

Goss, P. J. E. and J. Peccoud (1999). Analysis of the stabilizing effect of ROM on the genetic network controlling ColE1 plasmid replication. *Pacific Symposium on Biocomputing.*

Griffith, J. S. (1968). "Mathematics of cellular control processes I. Negative feedback to one gene." *Journal of Theoretical Biology* 20: 202-208.

Griffith, J. S. (1968). "Mathematics of cellular control processes II. Positive feedback to one gene." *Journal of Theoretical Biology* 20: 209-216.

Gui, J., A. S. Andrew, et al. (2011). "A Robust Multifactor Dimensionality Reduction Method for Detecting Gene–Gene Interactions with Application to the Genetic Analysis of Bladder Cancer Susceptibility." *Annals of Human Genetics* 75: 20-28.

Hartemink, A. J., D. K. Gifford, et al. (2001). Using graphical models and genomic expression data to statistically validate models of genetic regulatory networks. *Pacific Symposium on Biocomputing.*

He, H., W. Oetting, et al. (2009`). "Pair-wise multifactor dimensionality reduction method to detect gene-gene interactions in a case-control study." *Human Heredity* 69(1): 60-70.

Heidema, A., J. Boer, et al. (2006). "The challenge for genetic epidemiologiests: how to analyze large numbers of SNPs in relation to complex diseases." *BMC Genetics* 7(23): doi: 10.1186/1471-2156-1187-1123.

Herwig, R., A. J. Poustka, et al. (1999). "Large-scale clustering of cDNA-fingerprinting data." *Genome Research* 9: 1093-1105.

Ideker, T., T. Galitski, et al. (2001). "A new approach to decoding life: system biology." *Annu. Rev. Genomics Hum. Genet.* 2: 343-372.

Ideker, T., V. Thorsson, et al. (2001). "Integrated genomic and proteomic analysis of a systematically perturbed metabolic network." *Science* 292: 929-934.

Jaffe, A., A. Feinberg, et al. (2011). "*Significance analysis and statistical dissection of variably methylated regions.*" Biostatistics.

Kitano, H. (2002). "Systems Biology: A Brief Overview, , March 1, 2002." *Science* 295: 1662-1664.

Koza, J., W. Mydlowec, et al. (2001). Reverse engineering of metabolic pathways from observed data using genetic programming. *Pacific Symposium on Biocomputing.*

Kraft, P. and D. Hunter (2010). The challenge of assessing complex gene-environment and gene-gene interactions. In Human genome epidemiology. Human genome epidemiology: building the evidence for using genetic information to improve health and prevent disease. M. J. Khoury, Oxford University Press 165-187.

Liang, S., S. Fuhrman, et al. (1998). REVEAL, A general reverse engineering algorithm for inference of genetic network architectures. *Pacific Symposium on Bio*computing.

Lin, Y., S. Lin, et al. (2009). "A gene expression signature that predicts the therapeutic response of the basal-like breast cancer to neoadjuvant chemotherapy." *Breast Cancer Research and Treatment* 123(3): 691-699.

Lucas, P. J. F., N. C. d. Bruijn, et al. (2000). "A probabilistic and decision-theoretic approach to the management of infectious disease at the ICU." *Artificial Intelligence in Medicine* 19(3): 251-279.

Matsuno, H., A. Doi, et al. (2000). Hybrid Petri net representation of gene regulatory network. *Pacific Symposium on Biocomputing.*

McAdams, H. and L. Shapiro (1995). "Circuit Simulation of Genetic Networks." *Science* 269(4): 650-656.

Michaels, G. S., D. B. Carr, et al. (1998). Cluster analysis and data visualization of large-scale gene expression data. Pacific Symposium on Biocomputing.

Murphy, K. and S. Mian (1999). Modelling Gene Expression Data using Dynamic Bayesian Networks. Technical report. U. B. Department of Computer Science.

Pearl, J. (1988). Probabilistic Reasoning in Intelligent Systems. San Mateo, CA, Morgan Kaufmann.

Pedraza, V., J. Gomez-Capilla, et al. (2010). "Gene expression signatures in breast cancer distinguish phenotype characteristics, histologic subtypes, and tumor invasiveness." *Cancer* 116(2): 486-496.

Quaglini, S., M. Stefanelli, et al. (2001). "Flexible guideline-based patient careflow systems." *Artificial Intelligence in Medicine* 22(1): 65-80.

Schwender, H. and K. Ickstadt (2007). "Identification of SNP interactions using logic regression." Oxford Journals.

Smith, E. N. and L. Kruglyak (2008). "Gene–Environment Interaction in Yeast Gene Expression." *PLoS Biology* 6(4): 810-824.

Somogyi, R. and C. Sniegoski (1996). "Modeling the complexity of genetic networks: understanding multigenetic and pleiotropic regulation." *Complexity* 1(6): 45-63.

Spellman, P. T., G. Sherlock, et al. (1998). "Comprehensive Identification of Cell Cycle-regulated Genes of the Yeast Saccharomyces cerevisiae by Microarray Hybridization." *Molecular Biology of the Cell* 9: 3273-3297.

Spirtes, P., C. Glymour, et al. (2001). Constructing Bayesian network models of gene expression networks from microarray data. to appear in the *Proceedings of the Atlantic Symposium on Computational Biology, Genome Information Systems and Technology.*

Tomita, M., K. Hashimoto, et al. (1999). "E-CELL: Software environment for whole cell simulation." *Bioinformatics* 15(1): 72-84.

Weaver, D. C., C. T. Workman, et al. (1999). Modeling regulatory networks with weight matrices. *Pacific Symposium on Biocomputing*, 123.

Wu, G., N. Yi, et al. (2011). "Statistical Quantification of Methylation Levels by Next-Generation Sequencing." *PLoS ONE* 6(6): e21034.

Yoo, C. (2010). Discovering gene-gene and gene-environment causal interactions using Bioinformatics approaches. Environmental Factors, Genes, and the Development of Human Cancers, Springer: 115-138.

Yoo, C., E. Blitz, et al. (2011). "Gene Pathways Discovery in Asbestos-Related Diseases using Local Causal Discovery Algorithm." Communications in Statistics - Simulation and Computation In Press.

Yoo, C. and G. Cooper (2004). "An Evaluation of a System that Recommends Microarray Experiments to Perform to Discover Gene-Regulation Pathways." *Journal of Artificial Intelligence in Medicine* 31: 169-182.

Yuh, C., H. Bolouri, et al. (1998). "Genomic Cis-regulatory logic: experimental and computational analysis of a sea urchin gene." *Science* 279: 1896-1902.

In: Bioinformatics Research: New Developments
Editors: C. Battik and K. Belhassine
ISBN: 978-1-61942-363-3

Chapter IV

Positive Selection, Putative Recombination Events and Evolutionary Historical Relationships among Members of the Family *Betaflexiviridae*

Moncef Boulila*
Institut de l'Olivier, Sousse Ibn-khaldoun, Tunisia

Abstract

Genetic variation of viruses is shaped by two major evolutionary forces: recombination and positive selection. In identifying putative recombination events in the complete genome of currently the known members of the family *Betaflexiviridae*, comprising 91 accessions retrieved from the international databases, in-depth molecular investigations were conducted by using RDP version 3.31β algorithm. It was demonstrated that potential recombination events were detected in three out of six virus genera (i.e, *Capillovirus*, *Foveavirus*, *Carlavirus*) along with the isolate TvAQ7 of the vitivirus *Grapevine virus E* and the unassigned *Betaflexiviridae*: *Sugarcane striate mosaic-associated virus*. Moreover, based on Fisher's Exact Test of Neutrality, positive selection operating on amino acid sequence levels of all virus genera except *Foveavirus* genus and the isolate DMV-932 of the citrivirus *Dweet mottle virus,* was also detected. It is worth noting that recombination and positive selection were both identified in representatives of *Capillovirus* and *Carlavirus* genera together with the unassigned *Betaflexiviridae*: *Sugarcane striate mosaic-associated virus*. The evolutionary history of all members of the family *Betaflexiviridae* showed that they segregated into seven distinct clusters corresponding to the six described virus genera and the clade "Unassigned *Betaflexiviridae*". From the present phylogenetic study, at least two

* Corresponding author: Moncef Boulila, Institut de l'Olivier, B.P. 14, 4061 Sousse Ibn-Khaldoun, Tunisia. phone/fax: 00216 73236135, E-mail address: boulila.moncef@ yahoo.fr.

taxonomic implications were drawn: (i) *Passiflora latent virus* should be considered as a typical carlavirus instead of belonging to the grouping "Unclassified carlaviruses". (ii) *Cherry virus A*, *Potato virus T*, *Grapevine virus E* and *Fig latent virus* should be extracted from the current groupings as actually stated by reports of the International Committee on Taxonomy of Viruses and gathered into a coherent ensemble to which the name "Unclassified *Betaflexiviridae*" should be allocated.

Keywords: Bioinformatics; Phylogeny; Recombination; natural selection; *Betaflexiviridae*; Sequence; Taxonomy

Introduction

Recombination is an important evolutionary force in many organisms and can have a profound impact on evolutionary rates. Viral RNA recombination, a process that joins together two noncontiguous RNA segments, is an especially powerful tool in virus evolution, since it can rapidly lead to dramatic changes in virus genomes by recombining or rearranging sequences. Accordingly, the significant role of RNA recombination in emergence of new viruses or virus strains is well documented for numerous viruses of different hosts (human, animal, plant, insect, fungus, bacterium) (Strauss *et al.*, 1988; Weiss *et al.*, 1991; Hajjou *et al.*, 1996; Desport *et al.*, 1998; Ayllon *et al.*, 1999). RNA recombination can be an efficient tool for viruses to repair viral genomes, thus contributing to virus fitness. It may also play a role in the formation of subviral RNAs that include defective interfering (DI) RNAs associated with many animal and plants viruses (Pathak and Nagy, 2009) as well as with chimeric satellite RNAs (Cheng and Nagy, 2003). It has been shown that RNA recombination enables the exchange of genetic material not only between the same or similar viruses but also between distinctly different viruses (Worobey and Holmes, 1999). Sometimes, it also permits crossovers between viral and host RNA (Greene and Allison, 1994; Aaziz and Tepfer, 1999, Baroth *et al.*, 2000, Nagai *et al.*, 2003). Taking into account the structure of viral genomic molecules and the location of crossover sites, three basic types of RNA recombination were distinguished: homologous, aberrant homologous and non-homologous (Lai, 1992; Alejska *et al.*, 2001). The former two occur between two identical or similar RNAs (or between molecules displaying local homology), while the latter involves two different molecules. Most of the collected data suggest that RNA recombinants are formed according to a copy choice model (Alejska *et al.,* 2001). A viral replication complex starts nascent RNA strand synthesis on one template, called RNA donor and then switches to another template, called RNA acceptor. Accordingly, two main factors are thought to affect RNA recombination: the structure of recombining segments and the ability of the viral replicase to switch templates.

RNA virus exists in a quasispecies nature, which represents microheterogeneity among viral population. The mutations occur associated with the lack of proofreading activity of the RNA polymerase during replication. The genome changes may render the viral genome able to replicate more efficiently or to escape the host immune responses (Wang *et al.*, 2007). Positive selection usually preserves such favorable mutations. In contrast, negative selection removes deleterious substitutions from the viral populations, thereby ensuring survival of the viral species. Through generations, viral populations evolve under various selective forces at different regions and sites that display different functional constraints. A stringent and robust criterion for detecting adaptive evolution in a protein-coding gene is an accelerated nonsynonymous (d_N, amino acid replacing) rate relative to the synonymous (d_S, silent) rate of substitutions, with the rate ratio $\omega = d_N/d_S > 1$. As silent mutations do not change the amino acid whereas replacement mutations do, the difference in their fixation rates provides a measure of selective pressure on the protein.

Both recombination and selective constraints can influence the evolutionary relationships among different viruses (Massey *et al.*, 2008; Boulila, 2009a,b; 2010). Revealing the evolutionary history of the members of RNA virus families has become one of the main goals for evolutionary virologists (Costa-Mattioli *et al.*, 2002; Katayama *et al.*, 2002; Lukashov and Goudsmit, 2002; de Villiers *et al.*, 2004; Thiery *et al.*, 2004), and the proposals for classification have been changing constantly in recent years. Many causes of this change are: (i) the lack of complete genome sequences in databanks leading to focusing on the analysis of a single gene, (ii) the presence of extensive recombination, (iii) the poor reliability of alignments due to the typically high sequence divergence of viruses and the existence of deletions and insertions. However, new techniques have been recently developed for analyzing unaligned multiple whole genomes and for producing global phylogenies (Stuart *et al.*, 2002a,b; Stuart and Berry, 2003; Qi *et al.*, 2004a,b).

Amongst positive-strand plant RNA viruses, the family *Betaflexiviridae* encompasses several viruses having an important economical impact. It includes the following genera: *Carlavirus*, *Citrivirus*, *Vitivirus*, *Trichovirus*, *Capillovirus*, and *Foveavirus*. In addition, a supplementary group: Unassigned *Betaflexiviridae* is also included in this family. The chief characteristics of members of the family are: (i) flexuous filamentous not enveloped virions, 610-960 nm long and 12-13 nm in diameter, (ii) monopartite, positive sense, single-stranded RNA genomes with a 3'-poly(A) tail, (iii) translation of at least some ORFs from subgenomic mRNAs (viruses in the family employ both 5'- and 3'-coterminal subgenomic mRNAs, (iv) up to six ORFs ordered from 5' to 3', (v) viruses of the genera *Capillovirus*, *Citrivirus*, *Trichovirus*, and *Vitivirus* contain a single movement protein (MP) of the "30K" superfamily (Melcher, 2000) and viruses of the genera *Carlavirus*, and *Foveavirus* encode a triple gene block (TGB) (Morozov and Solovyev, 2003), (vi) a single capsid (coat) protein (CP) (21-44 kDa), (vii) a replication protein (190-250 kDa) containing conserved methyl transferase, helicase and RNA-dependent RNA polymerase (RdRp) motifs (Koonin and Dolja, 1993). Biologically, the viruses are fairly diverse. They have been reported from a wide range of herbaceous and woody mono- and dicotyledonous plant species but the host range of individual members is usually limited. Natural infections by components of the genera *Foveavirus*, *Capillovirus*, *Vitivirus*, and *Trichovirus* are mostly or exclusively of woody hosts. Many of the viruses have relatively mild effects upon their host. Major species can be transmitted by mechanical inoculation, often readily. Many of the viruses have no known invertebrate or fungal vectors but some trichoviruses are thought to be mite-borne, most carlaviruses are transmitted naturally by aphids in the persistent manner and a range of vectors have been reported for different vitiviruses. Aggregates of virus particles accumulate in the cytoplasm (Adams *et al.*, 2004).

Table 1. Current taxonomic status of the members of the family *Betaflexiviridae* included in the study and their accession numbers

Genus (in bold), unclassified and unassigned viruses	Virus.Isolate	GenBank accession number
Capillovirus	(Apple stem grooving virus) (ASGV)	NC_001749
	(Apple stem grooving virus.Li-23) (ASGV.Li-23)	AB004063
	(Cherry virus A) (CVA)	NC_003689
	(Citrus tatter leaf virus.Lily) (CTLV.Lily)	D16681
	(Citrus tatter leaf virus.Kumquat 1) (CTLV. Kumquat 1)	AY646511
	(Citrus tatter leaf virus.Lcd-NA-1) (CTLV.Lcd-NA-1)	FJ355920
	(Citrus tatter leaf virus.Meyer lemon) (CTLV.Meyer lemon)	EU553489
	(Pear black necrotic leaf spot virus.South Korea) (PBNLSV.South Korea)	AY596172
Citrivirus	(Dweet mottle virus.DMV-932) (DMV-932)	FJ009367
Foveavirus	(Apple stem pitting virus.PA66) (ASPV.PA66)	NC_003462
	(Apple stem pitting virus.PR1) (ASPV.PR1)	EU095327
	(Peach chlorotic mottle virus.Agua-4N6) (PCMoV.Agua-4N6)	NC_009892
	(Rupestris stem pitting-associated virus 1) (RSPaV 1)	AF057136
	(Rupestris stem pitting-associated virus) (RSPaV)	AF026278
	(Rupestris stem pitting-associated virus.BS) (RSPaV.BS)	AY881627
	(Rupestris stem pitting-associated virus.PN) (RSPaV.PN)	AY368172
	(Rupestris stem pitting-associated virus.SG1) (RSPaV.SG1)	AY881626
Trichovirus	Apple chlorotic leaf spot virus. TA TAO 5 (ACLSV.TA TAO 5)	EU223295
	Apple chlorotic leaf spot virus.A4 (ACLSV.A4)	AB326223
	Apple chlorotic leaf spot virus.P-205 (ACLSV.P-205)	D14996
	Apple chlorotic leaf spot virus.B6 (ACLSV.B6)	AB326224
	Apple chlorotic leaf spot virus. Balaton 1(ACLSV.Balaton 1)	X99752
	Apple chlorotic leaf spot virus.MO-5 (ACLSV.MO-5)	AB326225
	Apple chlorotic leaf spot virus.P863 (ACLSV.P863)	M58152
	Apple chlorotic leaf spot virus.PBM1 (ACLSV.PBM1)	AJ243438
Unclassified trichoviruses	(Apricot pseudo-chlorotic leaf spot virus.Sus2) (APCLSV.Sus 2)	NC_006946
	(Fig latent virus 1.f5p5) (FLV 1.f5p5)	FN377573
Vitivirus	Grapevine virus A (GVA)	AY244516
	Grapevine virus A.BMo 32-1 (GVA.BMo 32-1)	DQ855087
	Grapevine virus A. GTG11-1 (GVA.GTG11-1)	DQ855084
	Grapevine virus A.GTRI-1 (GVA.GTRI-1)	DQ787959
	Grapevine virus A.GTRI-2 (GVA.GTRI-2)	DQ855086
	Grapevine virus A.GTR1SD-1 (GVA.GTR1SD-1)	DQ855081
	Grapevine virus A.Is 151 (GVA.Is 151)	NC_003604
	Grapevine virus A.KWVMo4-1 (GVA.KWVMo4-1)	DQ855083
Vitivirus	Grapevine virus A.P163-1 (GVA.P163-1)	DQ855088
	Grapevine virus A.P163-M5 (GVA.P163-M5)	DQ855082
	Grapevine virus A.PA3 (GVA.PA3)	AF007415
	Grapevine virus B (GVB)	NC_003602
	Grapevine virus B.94.971 (GVB.94.971)	EF583906
	Grapevine virus E.TvAQ7 (GVE.TvAQ7)	AB432910
Carlavirus	(Aconitum latent virus.D) (ALV.D)	NC_002795
	(Blueberry scorch virus.BC-1) (BSV.BC-1)	AY941198
	(Blueberry scorch virus.BC-2) (BSV.BC-2)	AY941199
	(Blueberry scorch virus.NJ-2) (BSV.NJ-2)	L25658
	(Chrysanthemum virus B.Punjab) (CVB.Punjab)	NC_009087
	(Chrysanthemum virus B.Uttar Pradesh) (CVB.Uttar Pradesh)	AM765837

Genus (in bold), unclassified and unassigned viruses	Virus.Isolate	GenBank accession number
	(Chrysanthemum virus B.S) (CVB.S)	AB245142
	(Chrysanthemum virus B.Tamil Nadu) (CVB. Tamil Nadu)	AM765839
	(Chrysanthemum virus B.Uttarakhand) (CVB.Uttarakhand)	AM765838
	(Coleus vein necrosis virus.Verbena) (CVNV.Verbena)	NC_009764
	(Daphne virus S.K) (DVS.K)	NC_008020
	(Garlic latent virus.Wonju) (GLV.Wonju)	Z68502
	(Garlic latent virus.YH1) (GLV.YH1)	NC_003557
	(Helleborus net necrosis virus.G5) (HNNV.G5)	NC_012038
	(Helleborus net necrosis virus.H6) (HNNV.H6)	FJ196836
	(Hop latent virus.Humulus lupulus cv.Shinshu–wase) (HLV.Shinshu-wase)	NC_002552
	(Hop mosaic virus.Humulus lupulus cv.Super Pride) (HMV.Super Pride)	NC_010538
	(Kalanchoe latent virus.1452) (KLV.1452)	FJ531635
	(Kalanchoe latent virus.PV-0290B) (KLV.PV.0290B)	NC_013006
	(Ligustrum necrotic ringspot virus) (LNRSV)	NC_010305
	(Lily symptomless virus) (LSV)	NC_005138
	(Lily symptomless virus.India) (LSV.India)	AM422452
	(Lily symptomless virus.Sb) (LSV.Sb)	AJ564638
	(Lily symptomless virus.SMi) (LSV.SMi)	AM263208
	(Narcissus common latent virus.Zhangzhou) (NCLV.Zhangzhou)	NC_008266
	(Potato virus M.DSMZ PV-0273) (PVM.DSMZ PV-0273)	EU604672
	(Potato virus M.Hangzhou) (PVM.Hangzhou)	AJ437481
	(Potato virus M.M57) (PVM.M57)	AY311395
	(Potato virus M. Russian wild) (PVM.Russian wild)	NC_001361
	(Potato virus M.Uran) (PVM.Uran)	AY311394
	(Potato virus S.Id4106-US) (PVS.Id4106-US)	FJ813513
	(Potato virus S.Leona) (PVS.Leona)	NC_007289
	(Potato virus S.Vltava) (PVS.Vltava)	AJ863510
	(Potato virus S.Wadef-US) (PVS.Wadef-US)	FJ813512
Unclassified carlaviruses	(Red clover mosaic virus.Washington) (RCMV.Washington)	NC_012210
	(Butterbur mosaic virus) (BuMV)	NC_013527
	(Hippeastrum latent virus) (HipLV)	NC_011540
	(Hydrangea chlorotic mottle virus.NZ) (HCMoV.NZ)	NC_012869
	(Narcissus symptomless virus.Hangzhou 2005) (NSV.Hangzhou 2005)	NC_008552
	(Passiflora latent carlavirus) (PLV)	NC_008292
	(Phlox virus B.WP) (PhVB.WP)	NC_009991
	(Phlox virus S.BR) (PhVS.BR)	NC_009383
	(Potato virus P.Bra) (PV P.Bra)	EU338239
	(Sweet potato chlorotic fleck virus) (SPCFkV)	NC_006550
Unassigned *Betaflexiviridae*	(African oil palm ringspot virus) (AOPRSV)	AY072921
	(Cherry necrotic rusty mottle virus) (CNRMoV)	NC_002468
	(Cherry necrotic rusty mottle virus.FC4) (CNRMoV.FC4)	EU188438
	(Cherry necrotic rusty mottle virus.FC5) (CNRMoV.FC5)	EU188439
	(Potato virus T) (PVT)	NC_011062
	(Sugarcane striate mosaic-associated virus) (SSMaV)	NC_003870

The objective of the present study was to investigate putative recombination events and selective constraints in the complete genome of currently the known members of the family *Betaflexiviridae* and finally to propose a reclassification based on the predicted evolutionary relationships among them. This was achieved by analyzing the genome of 91 accessions retrieved from GenBank.

Materials and Methods

Virus Sequence Source

The sequences of the full-length genome of 91 accessions used in this study were downloaded from GenBank (Table 1).

Sequence Alignments, Selective Constraints, Recombination and Phylogenetic Analyses

The nucleotide sequences were aligned using CLUSTALW 2.0.9 and CLUSTALX 2.0.9 (Larkin *et al.,* 2007) with default parameters.

Their phylogenetic relationships were determined by five methods using the neighbor-joining (NJ), Minimum Evolution (ME), Maximum Parsimony (MP), and Unweighted Pair Group Method with Arithmetic mean (UPGMA) implemented in MEGA4.1β program (Kumar *et al.,* 2008) and Maximum-likelihood (ML) using PHYLO_WIN graphic tool (Galtier *et al.*, 1996). Bootstrap analyses with 500 replicates were performed to assess the robustness of the branches.

Positive selection, using MEGA4.1β program, was inferred by the counting method described by Nei and Gojobori (1986) and later by Suzuki and Gojobori (1999). This method utilized the phylogenetic tree of sequences analyzed. In the parsimony method, the total numbers of synonymous (c_S) and nonsynonymous (c_N) substitutions as well as the average numbers of synonymous (s_S) and nonsynonymous (s_N) sites per codon over the phylogenetic tree for each codon site were computed according to the maximum parsimony principle (Fitch, 1971; Hartigan, 1973). The null hypothesis of selective neutrality (r_S=r_N or $\omega = 1$) was tested for each site by computing the probability (p) of obtaining the observed or more biased values for c_S and c_N, which were assumed to follow a binomial distribution with the probabilities of occurrence of synonymous and nonsynonymous substitutions given by $s_S/(s_S + s_N)$ and $s_N/(s_S + s_N)$, respectively. Positive selection is inferred when p<0.05 and $c_N/s_N > c_S/s_S$ (Suzuki, 2006).

Potential recombination events between diverged nucleotide sequences were explored with the program RDP v3.31β (Martin *et al.,* 2005b). RDP incorporates several published recombination detection methods into a single suite of tools: RDP (Martin and Rybicki, 2000), GENECONV (Padidam *et al.,* 1999), BOOTSCAN (Martin *et al.*, 2005a), MAXCHI (Smith, 1992), CHIMAERA (Posada and Crandall, 2001), SISCAN (Gibbs *et al.,* 2000), and 3SEQ (Boni *et al.,* 2007). In all cases, defaults parameters were used. Only events predicted by more than half of the methods are considered as significant.

Table 2. Determination of inferred putative recombination events in the entire genome of members of the virus genera: *Capillovirus*, *Foveavirus*, *Vitivirus*, *Carlavirus* and Unassigned *Betaflexiviridae*

Recombinant virus.isolate	Accession number	Position of the recombining fragment in the sequence (nt)	Position of the recombining fragment in the alignment (nt)	Accession number of putative parentals		Putative parentals (major x minor)
				Major parent	Minor parent	
ASGV	NC_001749	3651-4334	5927-6617	DQ855081	AY368172	GVA.GTR1SD-1 x RSPaV.PN
ASGV.Li-23	AB004063	3643-3976	5919-6258	DQ855081	AY368172	GVA.GTR1SD-1 xRSPaV.PN
		4342-4554	6625-6847	EU553489	AY596172	CTLV.Meyer lemon x PBNLSV
		5849-6189	8493-8959	AY596172	FJ355920	PBNLSV.S. Korea x CTLV.Lcd-NA-1
CTLV.Lily	D16681	3665-3976	5941-6258	DQ855081	AY368172	GVA.GTR1SD-1 x RSPaV.PN
		4342-4554	6625-6847	EU553489	AY596172	CTLV.Meyer lemon x PBNLSV
		5842-6189	8477-8959	AY596172	FJ355920	PBNLSV.S. Korea x CTLV.Lcd-NA-1
CTLV.Kumquat 1	AY646511	3645-4348	5921-6631	DQ855081	AY368172	GVA.GTR1SD-1 x RSPaV.PN
CTLV.Lcd-NA-1	FJ355920	3666-4348	5942-6631	DQ855081	AY368172	GVA.GTR1SD-1 x RSPaV.PN
CTLV.Meyer lemon	EU553489	3628-4348	5903-6630	DQ855081	AY368172	GVA.GTR1SD-1 x RSPaV.PN
PBNLSV.S. Korea	AY596172	3667-3977	5942-6258	DQ855081	AY368172	GVA.GTR1SD-1 x RSPaV.PN
ASPV.PR1	EU095327	6017-6477	6493-6975	EU188438	NC_008266	CNRMoV.FC4 x NCLV.Zhangzhou
PCMV.Agua 4N6	NC_009892	4917-5474	5477-6068	EU095327	NC_012210	ASPV.PR1 x RCMV.Washington
RSPaV 1	AF057136	6071-6442	6570-6960	EU095327	NC_009383	ASPV.PR1 x PhVS.BR
RSPaV	AF026278	6056-6452	6556-6974	EU095327	NC_009383	ASPV.PR1 x PhVS.BR
RSPaV.BS	AY881627	6035-6462	6535-6984	EU095327	NC_009383	ASPV.PR1 x PhVS.BR
		7527-8707	8284-9829	AY881626	AY368172	RSPaV.SG1 x RSPaV.PN
RSPaV.PN	AY368172	6275-7421	6794-8117	AF026278	AY881627	RSPaV x RSPaV.BS
RSPaV.SG1	AY881626	6044-6504	6544-7026	EU095327	NC_009383	ASPV.PR1 x PhVS.BR
GVE.TvAQ7	AB432910	6805-7564	9040-10290	AB004063	DQ855086	ASGV.Li-23 x GVA.GTRI-2
BSV.BC-1	AY941198	7389-8126	8824-9660	NC_009087	AY881626	CVB.Punjab x RSPaV.SG1
BSV.BC-2	AY941199	7387-8129	8819-9660	NC_009087	AY881626	CVB.Punjab x RSPaV.SG1
BSV.NJ-2	L25658	7515-8120	8966-9660	NC_009087	AY881626	CVB.Punjab x RSPaV.SG1
CVB.Punjab	NC_009087	577-786	720-996	AB245142	AY941199	CVB.S x BSV.BC-2
		2196-2547	2553-3077	EU338239	NC_012869	PVP.Bra x HCMoV.NZ
		4292-5275	5035-6074	AM765837	AB245142	CVB.Uttar Pradesh x CVB.S
		5276-6983	6075-7965	AM765837	AM765838	CVB.Uttar Pradesh x CVB.Uttarkhand
		6987-7434	7969-8492	AM765838	AM765839	CVB.Uttarkhand x CVB.Tamil Nadu
		8466-end	9661-end	AM765838	AM765839	CVB.Uttarkhand x CVB.Tamil Nadu
CVB.Uttar Pradesh	AM765837	577-785	720-995	AB245142	AY941199	CVB.S x BSV.BC-2
		2187-2538	2553-3077	EU338239	NC_012869	PVP.Bra x HCMoV.NZ
		4280-4607	5032-5376	AM765838	AB245142	CVB.Uttarkhand x CVB.S
		4871-5269	5652-6077	AM765838	AB245142	CVB.Uttarkhand x CVB.S
		6978-7425	7969-8492	AM765838	AM765839	CVB.Uttarkhand x CVB.Tamil Nadu
		8503-end	9721-end	AM765838	AM765839	CVB.Uttarkhand x CVB.Tamil Nadu

Table 2. (Continued)

Recombinant virus.isolate	Accession number	Position of the recombining fragment in the sequence (nt)	Position of the recombining fragment in the alignment (nt)	Accession number of putative parentals		Putative parentals (major x minor)
				Major parent	Minor parent	
CVB.S	AB245142	537-4365	616-5008	AM765839	AM765838	CVB.Tamil Nadu x CVB.Uttarkhand
CVB.Tamil Nadu	AM765839	674-3153	819-3375	AM765837	AM765838	CVB.Uttar Pradesh x CVB.Uttarkhand
CVB.Uttarakhand	AM765838	577-768 2184-2535	720-977 2553-3077	AB245142 EU338239	AY941199 NC_012869	CVB.S x BSV.BC-2 PVP.Bra x HCMoV.NZ
CVNV.Verbena	NC_009764	2360-2587 8335-end	2800-3229 9799-end	AJ863510 NC_003557	NC_013006 NC_001361	PVS.Vltava x KLV.PV.0290B GLV.YH1 x PVM.Russian wild
GLV.YH1	NC_003557	8105-8359	9817-10120	NC_001361	NC_008552	PVM.Russian wild x NSV.Hangzhou 2006
HNNV.G5	NC_012038	2019-2367	2508-2995	AJ863510	NC_013006	PVS.Vltava x KLV.PV.0290B
HNNV.H6	FJ196836	2017-2513	2506-3233	AJ863510	NC_013006	PVS.Vltava x KLV.PV.0290B
KLV.1452	FJ531635	7459-8134 8478-end	8890-9660 10085-end	NC_009087 L25658	AY881626 AB245142	CVB.Punjab x RSPaV.SG1 BSV.NJ-2 x CVB.S
KLV.PV.0290B	NC_013006	7376-end	8137-end	L25658	AB245142	BSV.NJ-2 x CVB.S
LNRSV	NC_010305	7291-8819	8033-9660	NC_009087	AY881626	CVB.Punjab x RSPaV.SG1
LSV	NC_005138	7251-8010 8352-end	8799-9660 10078-end	NC_009087 L25658	AY881626 AB245142	CVB.Punjab x RSPaV.SG1 BSV.NJ-2 x CVB.S
LSV.India	AM422452	7251-8010 8371-end	8799-9660 10100-end	NC_009087 L25658	AY881626 AB245142	CVB.Punjab x RSPaV.SG1 BSV.NJ-2 x CVB.S
LSV.Sb	AJ564638	7251-8010	8799-9660	NC_009087	AY881626	CVB.Punjab x RSPaV.SG1
LSV.SMi	AM263208	7369-8010	8929-9660	NC_009087	AY881626	CVB.Punjab x RSPaV.SG1
PVM.DSMZ.PV0273	EU604672	2191-2602	2377-2916	NC_010538	NC_009764	HMV.Sup. Pride x CVNV.Verbena
PVM.Hangzhou	AJ437481	2193-2377	2604-2916	NC_010538	NC_009764	HMV.Sup. Pride x CVNV.Verbena
PVM.Russian wild	NC_001361	2191-2377	2602-2916	NC_010538	NC_009764	HMV.Sup. Pride x CVNV.Verbena
PVM.Uran	AY311394	2189-2375	2602-2916	NC_010538	NC_009764	HMV.Sup. Pride x CVNV.Verbena
PVM.M57	AY311395	2189-2375	2602-2916	NC_010538	NC_009764	HMV.Super Pride x CVNV.Verbena
PVS.Id4106-US	FJ813513	2346-2564 7360-8095	2800-3225 8824-9642	NC_010305 NC_009087	NC_003557 AY881626	LNRSV x GLV.YH1 CVB.Punjab x RSPaV.SG1
PVS.Leona	NC_007289	2342-2560 7353-8824	2800-3225 8090-9660	NC_010305 NC_009087	NC_003557 AY881626	LNRSV x GLV.YH1 CVB.Punjab x RSPaV.SG1
PVS.Vltava	AJ863510	2339-2561 7298-8077	2802-3232 8743-9642	NC_010305 NC_009087	NC_003557 AY881626	LNRSV x GLV.YH1 CVB.Punjab x RSPaV.SG1
PVS.Wadef-US	FJ813512	2346-2564 7360-8097	2800-3225 8824-9660	NC_010305 NC_009087	NC_003557 AY881626	LNRSV x GLV.YH1 CVB.Punjab x RSPaV.SG1
HCMoV.NZ	NC_012869	7383-8035	8918-9660	NC_009087	AY881626	CVB.Punjab x RSPaV.SG1

Recombinant virus.isolate	Accession number	Position of the recombining fragment in the sequence (nt)	Position of the recombining fragment in the alignment (nt)	Accession number of putative parentals		Putative parentals (major x minor)
				Major parent	Minor parent	
PLV	NC_008292	7395-8000	8966-9660	NC_009087	AY881626	CVB.Punjab x RSPaV.SG1
		8381-end	10119-end	L25658	AB245142	BSV.NJ-2 x CVB.S
PhVB.WP	NC_009991	2194-2615	2600-3231	NC_011540	NC_010538	HipLV x HMV.Super pride
		8567-8852	9830-10121	NC_009383	L25658	PVS.BR x BSV.NJ-2
PhVS.BR	NC_009383	2094-2477	2499-3070	EU338239	NC_012869	PVP.Bra x HCMoV.NZ
PVP.Bra	EU338239	7263-7997	8824-9660	NC_009087	AY881626	CVB.Punjab x RSPaV.SG1
SSMaV	NC_003870	5635-5850	6826-7044	NC_003462	AM263208	ASPV.PA66 x LSV.SMi

Table 3. Accession pairs of the *Betaflexiviridae* under positive selection with determined probability at the 5% level based on the Fisher's Exact Test of Neutrality and calculated d_N-d_S

Accession pairs	Virus.isolate pairs	*P* value at 5% level	d_N-d_S
NC_003689/NC_003870	CVA/SSMaV	0.037	1.915
NC_003602/FN377573	GVB/FLV 1.f5p5	0.030	1.990
EF583906/X99752	GVB.94.971/ACLSV.Balaton 1	0.031	1.938
EF583906/ NC_003689	GVB.94.971/CVA	0.034	2.042
EF583906/ FN377573	GVB.94.971/FLV.f5p5	0.040	1.929
EF583906/M58152	GVB.94.971/ACLSV.P863	0.013	2.359
EF583906/AJ243438	GVB.94.971/ACLSV.PBM1	0.043	1.785
AJ564638/AY646511	LSV.Sb/CTLV.Kumquat 1	0.017	2.230
AJ564638/FJ355920	LSV.Sb/CTLV.Lcd.NA-1	0.017	2.451
AJ564638/EU553489	LSV.Sb/CTLV.Meyer Lemon	0.018	2.233
AM422452/ AY646511	LSV.India/CTLV.Kumquat 1	0.015	2.291
AM422452/ FJ355920	LSV.India/CTLV.Lcd.NA-1	0.011	2.550
AM422452/ EU553489	LSV.India/CTLV.Meyer lemon	0.019	2.245
AM263208/ AY646511	LSV.SMi/CTLV.Kumquat 1	0.012	2.405
AM263208/ FJ355920	LSV.SMi/CTLV.Lcd.NA-1	0.015	2.481
AM263208/ EU553489	LSV.SMi/CTLV.Meyer lemon	0.013	2.415
AY941198/ NC_003870	BSV.Bc-1/SSMaV	0.009	2.510
NC_005138/ AY646511	LSV/CTLV.Kumquat 1	0.022	2.210
NC_005138/ FJ355920	LSV/CTLV.Lcd.NA-1	0.016	2.468
NC_005138/ EU553489	LSV/CTLV.Meyer lemon	0.026	2.144
NC_008292/ NC_003870	PLV/SSMaV	0.007	2.770

Results

Recombination Events during Evolution of the Betaflexiviridae

Examination of RDP program package output showed that among capilloviruses, ASGV, PBNLSV, and the isolates Lily, Kumquat 1, Lcd-NA-1, and Meyer lemon of CTLV, showed putative breakpoints in their genome. Among foveaviruses, all investigated accessions were potential recombinants except the isolate ASPV.PA66. About 60% of members of *Carlavirus* genus (BSV, CVB, CVNV, GLV, HNNV, KLV, LNRSV, LSV, PVM, PVS, HCMoV, PLV, PhVS, PhVB, PVP) were possible recombinants. Similarly, the unassigned *Betaflexiviridae* (SSMaV) was also a potential recombinant. Furthermore, no possible recombination sites were detected in the genome of members of *Trichovirus*, *Citrivirus* and *Vitivirus* (except GVE.TvAQ7) genera. It is worth pointing out that detected putative recombination breakpoints consisted in one (70%), two (22%), three (4%) or six (4%) recombination sites distributed in the entire genome of all potential recombining accessions (Table 2).

Nucleotide Sequence Analysis

Maximum composite likelihood estimate of the pattern of nucleotide substitution were conducted in MEGA4.1β. The results for the *Betaflexiviridae* were as follows: rates of different transitional substitutions varied from 10.34 to 13.71, and those of transversional substitutions varied from 5.14 to 7.4. The nucleotide frequencies were as follows: 0.284 (A),

0.262 (T/U), 0.197 (C), and 0.256 (G). The transition/transversion rate ratios were $k_1 = 1.698$ (purines) and $k_2 = 2.012$ (pyrimidines). The overall transition/transversion bias was $R = 0.814$, where $R = [AGk_1 + TCk_2]/[(A+G)(T+C)]$. Codon positions included were first + second + third + noncoding. All positions containing gaps and missing data were eliminated from the dataset (complete deletion option). There were a total of 3543 positions in the final dataset.

The MEGA4.1β program incorporates also the Tajima's Neutrality Test. The purpose of this test is to indentify sequences which do not fit the neutral theory model at equilibrium between mutation and genetic drift. Tajima's test compares a standardized measure of the total number of segregating sites (these are the DNA sites that are polymorphic) in the sampled DNA and the average number of mutations between pairs in the sample. The Tajima's D was determined ($D = 6.440827$).

Positive Selection

The high genetic stability of viruses can be attributed to negative or purifying selection to maintain the functional integrity of the viral genome. The degree of negative selection in genes, or the degree of functional constraint for the maintenance of the encoded protein sequence, can be estimated, as indicated above, by the ratio between the nucleotide diversities in nonsynonymous and synonymous positions (d_N/d_S). For most coding genes the d_N/d_S ratio is below unity which is consistent with negative selection against protein change. In contrast, a d_N/d_S ratio above unity may be an indication that adaptive or positive selection is driving gene divergence. In this work, pairwise comparisons of all screened accessions showed that 21 accession pairs were under positive selection involving 17 different lineages (Table 3). Except *Citrivirus* and *Foveavirus*, all remaining virus genera contained viruses under positive selection. In fact, in *Capillovirus* genus, the following viruses were under positive selection: CVA, and CTLV (isolate Kumquat 1, isolate Lcd-NA-1, isolate Meyer lemon) representing a rate of 50% (4/8). In *Trichovirus* genus, the following isolates of ACLSV were under positive selection: Balaton 1, P836, PBM1, together with FLV (isolate f5p5) representing a rate of 40% (4/10). In *Vitivirus* genus, only GVB (unknown isolate and isolate 94.971) were under positive selection with a rate of 14% (2/14). In *Carlavirus* genus, the following viruses were under positive selection: BSV (isolate BC-1), and LSV (unknown isolate, isolate India, isolate Sb, isolate SMi) together with PLV representing a rate of 13% (6/44). Only one unassigned *Betaflexiviridae* was under positive selection: SSMaV with a rate of 16% (1/6). All these results were obtained by testing the neutrality in sequence pairs with Fisher's Exact Test. The probability of rejecting the null hypothesis of strict-neutrality ($d_N = d_S$) in favor of positive selection for each sequence pair was determined. Values of P less than 0.05 were considered significant at the 5% level. The variance of the difference (d_N-d_S) was computed using the bootstrap method (500 replicates) (Table 3). All analyses were conducted using the Nei-Gojobori method incorporated in MEGA program. All positions containing gaps and missing data were eliminated from the dataset (complete deletion option). There were a total of 466 positions in the final dataset.

Phylogenetic Relationships

The phylogenetic relationships among members of the family *Betaflexiviridae*, based on the sequences of their complete genome, were inferred using NJ, ME, MP, ML, and UPGMA. The constructed trees basically gave a somewhat similar phylogenetic grouping for the five methods used in this study. As a general picture, the inferred phylogeny showed that each taxonomical genus in the family *Betaflexiviridae* constitutes a homogenous group clearly distinct from one another (Figure 1). However, the results obtained in this study pointed out a few discrepancies with the current classification provided so far by the International Committee on Taxonomy of Viruses (ICTV). The main differences reside in the following points: (i) CVA is genetically distant to the other components of the *Capillovirus* genus. (ii) GVE is phylogenetically clearly distinct from the other members of *Vitivirus* genus (GVA, GVB), (iii) FLV cannot be considered as a closer relative of currently the known members of *Trichovirus* genus even with the allocation "Unclassified trichovirus, (iv) PVT was shown to be distantly related to the other unassigned *Betaflexiviridae*.

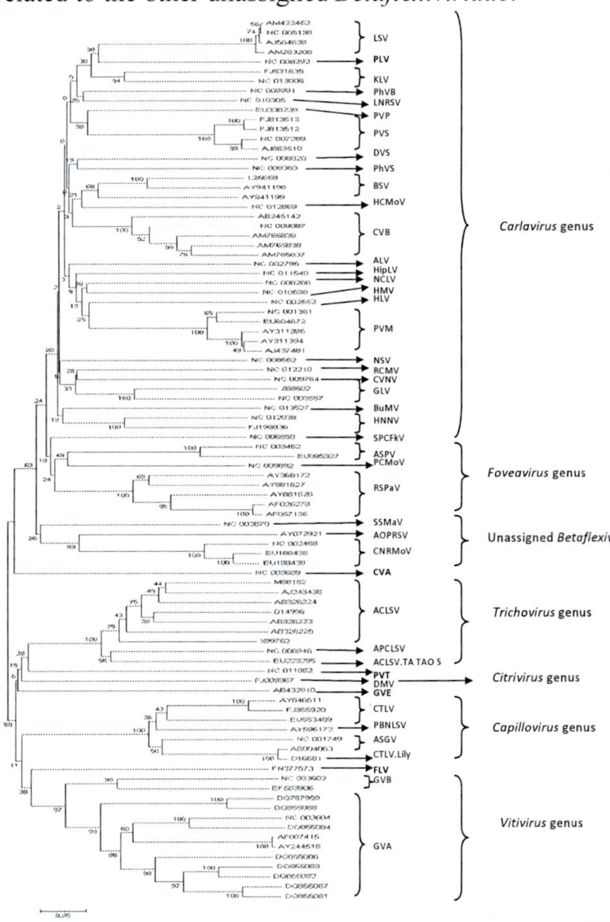

Figure 1. Dendrogram depicting phylogenetic relationships among the studied members of the family *Betaflexiviridae* based on the analysis of their complete genome sequences. Six clusters representing the six genera along with the grouping "Unassigned *Betaflexiviridae*" were clearly delineated. The tree was produced using the NJ algorithm option of MEGA4.1β. The numbers above the branches indicate the bootstrap confidence value. The scale bar shows the number of substitutions per nucleotide.

Finally, in the author's opinion, an update of the current classification should take place and consider that these four viruses would be seen as unclassified *Betaflexiviridae*. Moreover, inferred phylogeny showed that PLV was more closely related to the typical carlavirus LSV than to the current "Unclassified carlaviruses" (Table 1). This is strongly plausible based on the fact that both evolved with recombination and positive selection mechanisms.

Discussion

The screening of the selection pressures that have shaped genetic variation forms a major part of many studies of molecular evolution. A common approach to this problem involves estimating the rates of nonsynonymous (d_N) and synonymous (d_S) substitutions. Kimura (1977) was the first to point out that their comparison can be a powerful tool for studying natural selection. Estimates of d_N significantly different from d_S, provide convincing evidence for nonneutral evolution. This approach is attractive as it does not make any assumptions regarding the demographic history of the population, unlike many neutrality tests [e.g. Tajima's test (1989)] which compare estimates of effective population size obtained using different measures of genetic variation. Currently, there are three classes of methods to study selection: those that count the number of nonsynonymous and synonymous substitutions along the phylogeny (counting methods), those that assume a distribution of rates across sites and infer the rate at which individual sites evolve given this distribution (random effects models), and those that estimate the ratio of nonsynonymous to synonymous substitutions on a site-by-site basis (fixed effects models) (Kosakovsky Pond and Frost, 2005). The counting methods involve estimating the number of nonsynonymous and synonymous changes that have occurred at each codon throughout the evolutionary history of the sample by reconstructing the ancestral sequences using parsimony (Suzuki and Gojobori, 1999) or likelihood-based methods (Suzuki, 2004). These methods are attractive as they are computationally fast and hence can be applied to large data sets and do not involve making any assumptions regarding the distribution of rates across sites. Accordingly, the counting method was well adapted to this study which involved a large dataset (91 accessions). This study led to the identification of positive selection exerted on the protein expression of capilloviruses (CVA, CTLV), trichoviruses (ACLSV, FLV), vitiviruses (GVB), and carlaviruses (BSV, LSV, PLV). No positive selection was found to act on amino acid sites of a citrivirus (DMV-932) as well as in all components of *Foveavirus* genus. In addition, only one unassigned *Betaflexiviridae* was under positive selection: SSMaV.

Natural selection is usually the most powerful mechanism or process causing evolution to occur, however, it only selects among the existing variation already in a population. It does not create new genetic varieties or new combinations of varieties. One of the sources of those new combinations of genes is recombination. It is responsible for producing combinations not found in earlier generations. The attrition rate, which is a weeding-out process, doesn't have to be nearly so extreme as it does with mutation to see evolutionary changes survive. Even a slight attrition rate with recombination can produce a dramatic effect, whereas with mutation, a small attrition rate produces only a small effect. In order to determine the occurrence of putative recombination events in the genome of viruses belonging to the family *Betaflexiviridae*, a detailed investigation was conducted and demonstrated that several members of the genera *Capillovirus*, *Foveavirus*, and *Carlavirus* together with the vitivirus GVE.TvAQ7 and the unassigned *Betaflexiviridae* (SSMaV) were possible recombinants. In contrast, no potential recombination breakpoints were found in the genome of a citrivirus (DMV-932), all screened isolates of the vitiviruses GVA and GVB, and all investigated members of *Trichovirus* genus (ACLSV, APCLSV, FLV). This could be explained by the existence of strong constraints in their genome imposed by virus-vector and/or virus-host interactions. It is noteworthy that a few viruses (three isolates of CTLV, one isolate of BSV, four isolates of LSV, PLV and SSMaV) evolved under both driving forces (recombination and positive selection). A synergism might be occurring. Such a synergism between recombination and natural selection may have played a major role in Darwinian molecular evolution.

In inferring evolutionary relationships of homologous sequences of members of the family *Betaflexiviridae*, reconstructed trees showed that members of the six genera (*Capillovirus, Citrivirus, Foveavirus, Trichovirus, Vitivirus, Carlavirus*) along with "Unassigned *Betaflexiviridae"*, segregated into six clearly delineated groupings. In addition, the inferred phylogeny pointed out that CVA, GVE, FLV, and PVT should be extracted from the current groupings as actually stated by reports of the ICTV and gathered into a coherent ensemble to which the name "Unclassified *Betaflexiviridae*" should be allocated. Furthermore, PLV should be considered as a typical carlavirus since it is genetically too distant to current relatives which altogether form the cluster "Unclassified carlaviruses".

Finally, this article describes a detailed study on recombination potentially occurring in the entire genome of all currently the known members of the family *Betaflexiviridae,* positive selection acting at their protein expression level as well as their phylogenetic reconstruction. In addition, we proposed a new classification based on their predicted evolutionary history.

References

Aaziz R. and Tepfer M. 1999. Recombination in RNA viruses and in virus-resistant transgenic plants. *J. Gen. Virol.* 80, 1339-1346.

Adams M.J., Antoniw J.F., Bar-Joseph M., Brunt A.A., Candresse T., Foster G.D., Martelli G.P., Milne R.G., Zavriev S.K. and Fauquet C.M. 2004. The new plant virus family *Flexiviridae* and assessment of molecular criteria for species demarcation. *Arch. Virol.* 149, 1045-1060.

Alejska M., Kurzyniska-Kokorniak A., Broda M., Kierzek R. and Figlerowicz M. 2001. How RNA viruses exchange their genetic material. *Acta Bioch. Polon.* 48, 391-407.

Ayllon M.A., Lopez C., Navas-Castillo J., Mawassi M., Dawson D.O., Guerri J., Flores R. and Moreno P. 1999. New defective RNAs from *Citrus tristeza virus*: evidence for a replicase driven template switching mechanism in their generation. *J. Gen. Virol.* 80, 817-821.

Baroth M., Orlich M., Thiel H.J. and Becher P 2000. Insertion of cellular NEDD8 coding sequences in a pestivirus. *Virology* 278, 456-466.

Boni M.F., Posada D. and Feldman M.W. 2007. An exact nonparametric method for inferring mosaic structure in sequence triplets. *Genetics* 176, 1035-1047.

Boulila M. 2009a. Molecular evidence for recombination in *Prunus necrotic ringspot virus*. *Plant Mol. Biol. Rep.* 27, 189-198.

Boulila M. 2009b. Recombination structure and genetic relatedness among members of the family *Bromoviridae* based on their RNAs 1 and 2 sequence analyses. *Virus Genes* 38(3), 435-444.

Boulila M. 2010. Putative recombination events and evolutionary history of five economically important viruses of fruit trees based on the coat protein-encoding gene sequence analysis. *Biochem Genet.* 48, 357-375.

Costa-Mattioli M., Cristina J., Romero H., Perez-Bercof R., Casane D., Colina R., Garcia L., Vega I., Glikman G., Romanowsky V., Castello A., Nicand E., Gassin M., Billaudel S. and Ferre V. 2002. Molecular Evolution of Hepatitis A virus: a new classification based on the complete VP1 protein. *J. Virol.* 76, 9516-9525.

Cheng C.P. and Nagy P.D. 2003. Mechanism of RNA recombination in Carmo- and *Tombusviridae*; evidence for template switching by the RNA-dependant RNA polymerase in vitro. *J. Virol.* 77(22), 12033-12047.

Desport M., Collins M.E. and Brownlie J. 1998. Genomic instability in BVDV: an examination of the sequence and structural influences on RNA recombination. *Virology* 246, 352-361.

de Villiers E.M., Fauquet C., Broker T.R., Bernard H.U., zur Hausen H. 2004. Classification of papillomaviruses. *Virology* 324, 17-27.

Fitch W.N. 1971. Toward defining the course of evolution minimum change for a specific tree topology. *Syst. Zool.* 20, 406-416.

Galtier N., Gouy M. and Gautier C. 1996. SEAVIEW and PHYLO_WIN: two graphic tools for sequence alignment and molecular phylogeny. *CABIOS* 12(6), 543-548.

Gibbs M.J., Armstrong J.S. and Gibbs A.J. 2000. Sister-scanning: a Monte Carlo procedure for assessing signals in recombinant sequences. *Bioinformatics* 16, 573-582.

Greene A.E. and Allison R.F. 1994. Recombination between viral RNA and transgenic plants transcripts. *Science* 263, 1423-1425.

Hajjou M., Hill J.R., Subramaniam S.V., Hu J.Y. and Raju R. 1996. Non homologous RNA-RNA recombination events at the 3' nontranslated region of the sindbis virus genome: hotspots and utilization of nonviral sequences. *J. Virol.* 70, 5153-5164.

Hartigan J.A. 1973. Minimum mutation fits to a given tree. *Biometrics* 29, 53-63.

Katayama K., Shirato-Horikochi H., Kojima S., kageyama T., Oka T., Hoshino F.B., Fukushi S., Shinohara M., Uchida K., Suzuki Y., Gojobori T. and Takeda N. 2002. Phylogenetic analysis of the complete genome of 18 Norwalk-like viruses. *Virology* 299, 225-239.

Kimura M. 1977. Preponderance of synonymous changes as evidence for the neutral theory of molecular evolution. *Nature* 38, 515-524.

Koonin E.V. and Dolja V.V. 1993. Evolution and taxonomy of positive-strand RNA viruses: implication of comparative analysis of amino acid sequences. *Crit. Rev. Biochem. Mol. Biol.* 28, 375-430.

Kosakovsy Pond S.L. and Frost S.D.W. 2005. Not so different after all: a comparision of methods for detecting amino acid sites under selection. *Mol. Biol. Evol.* 22(5), 1208-1222.

Kumar S., Nei M., Dudley J. and Tamura K. 2008. MEGA: A biologist-centric software for evolutionary analysis of DNA and protein sequences. *Brief. Bioinfor.* 9(4), 299-306.

Lai M.M.C.. 1992. RNA recombination in animal and plant viruses. *Microbiol. Rev.* 56, 61-79.

Massey S.E., Churbanov A., Rastogi S. and Liberles D..A. 2008. Characterizing positive and negative selection and their phylogenetic effects. *Gene*, 418, 22-26.

Larkin M.A., Blackshileds G., Brown N.P., Chenna R., McGettigan P.A., McWilliam H., Valentin F., Wallace I.M., Wilm A., Lopez R., Thompson J.D., Gibson T.J. and Higgins D.G. 2007. Clustal W and Clustal X version 2.0. *Bioinformatics* 23(21), 2947-2948.

Lukashov V.V. and Goudsmit J. 2002. Evolutionary relationships among *Astriviridae*. *J. Gen. Virol.* 83, 1397-1405.

Martin D. and Rybicki E. 2000. RDP: detection of recombination amongst aligned sequences. *Bioinformatics* 16, 562-563.

Martin D.P., Posada D., Crandall K.A. and Williamson C. 2005a. A modified bootscan algorithm for automated identification of recombination sequences and recombination breakpoints. *AIDS Res. H. Retrovir.* 21, 98-102.

Martin D.P., Williamson C. and Posada D. 2005b. RDP2: recombination detection and analysis from sequence alignments. *Bioinformatics* 21, 260-262.

Melcher U. 2000. The '30K' superfamily of viral movement proteins. *J. Gen. Virol.* 81, 257-268.

Morozov S.Y. and Solovyev A.G. 2003. Triple gene block: modular design of a multifunctional machine for plant virus movement. *J. Gen. Virol.* 84, 1351-1366.

Nagai M., Sakoda Y., Mori M., Hayashi M., Kida H. and Akashi H. 2003. Insertion of a cellular sequence and RNA recombination in the structural protein coding region of cytopathogenic bovine viral diarrhea virus. *J. Gen. Virol.* 84Pt2, 447-452.

Nei M. and Gojobori T. 1986. Simple methods for estimating the numbers of synonymous and nonsynonymous nucleotide substitutions. *Mol. Biol. Evol.* 3(5), 418-426.

Padidam M., Sawyer S. and Fauquet C.M. 1999. Possible emergence of new geminiviruses by frequent recombination. *Virology* 265, 218-225.

Pathak K.B. and Nagy P.D. 2009. Defective interfering RNAs: foes of viruses and friends of virologists. *Viruses* 1, 895-919.

Posada D. Crandall K. 2001. Evaluation of methods for detecting recombination from DNA sequences. Computer simulation. *Proc. Natl. Acad. Sci. USA* 98, 13757-13762.

Qi J., Luo H. and Hao B. 2004a. CVTree: a phylogenetic tree reconstruction tool based on whole genomes. *Nucleic Acids Res.* 32, W45-W47.

Qi J., Wang B. and Hao B. 2004b. Whole genome prokaryote phylogeny without sequence alignment: a K-string composition approach. *J. Mol. Evol.* 58, 1-11.

Strauss J.H. and Strauss E.G. 1988. Evolution of RNA viruses. *Ann. Rev. Microbiol.* 42, 657-683.

Stuart G.W., Moffet K. and Baker S. 2002a. Integrated gene and species phylogenies from unaligned whole genome protein sequences. *Bioinformatics* 18, 100-108.

Stuart G.W., Moffet K. and Leader J.J. 2002b. A comprehensive vertebrate phylogeny using vector representations of proteins sequences from whole genome. *Mol. Biol. Evol.* 19, 554-562.

Stuart G.W. and Berry M.W. 2003. A comprehensive whole genome bacterial phylogeny using correlated peptide motifs defined in a high dimensional vector space. *J. Bioinfor. Comp. Biol.* 1, 475-493.

Suzuki Y. 2004. New methods for detecting positive selection at single amino acid sites. *J. Mol. Evol.* 59, 11-19.

Suzuki Y. 2006. Statistical properties of the methods for detecting positively selected amino acid sites. *Gene* 365, 125-129.

Suzuki Y. and Gojobori T. 1999. A method for detecting positive selection at single amino acid sites. *Mol. Biol. Evol.* 16(10), 1315-1328.

Tajima F. 1989. Statistical-method for testing the neutral mutation hypothesis by DNA polymorphism. *Genetics* 123, 585-595.

Thiéry R., Cozien J., de Boisséson C., Kerbart-Bosher S. and Névarez L. 2004. Genomic classification of new betanodavirus isolates by phylogenetic analysis of the coat protein suggests a low host-fish species specificity. *J. Gen. Virol.* 85, 3079-3087.

Wang S.Y., Wu J.C., Chiang T.Y., Huang Y.H., Su C.W. and Sheen I.J. 2007. Positive selection of Hepatitis delta antigen in chronic Hepatitis D patients. *J. Virol.* 81(9), 4438-4444.

Weiss B.G. and Schlesinger S. 1991. Recombination between Sindbis virus RNAs. *J. Virol.* 65 4017-4025.

Worobey M. and Holmes E.C. 1999. Evolutionary aspects of recombination in RNA viruses. *J. Gen. Virol.* 80, 2535-2543.

In: Bioinformatics Research: New Developments
Editors: C. Battik and K. Belhassine
ISBN: 978-1-61942-363-3

Chapter V

Selection, Recombination and Phylogenetics of Filamentous Plant RNA Viruses of the Family *Alphaflexiviridae*

Moncef Boulila*
Institut de l'Olivier, Sousse Ibn-khaldoun, Tunisia

Abstract

Positive selection and recombination are two major driving forces of virus evolution. In assessing potential recombination events and their frequency through the sequence of the whole genome of 73 accessions of the family *Alphaflexiviridae* retrieved from the international databases, a detailed study was conducted by using two algorithms: RECCO and RDP version 3.31β. These programs allowed the detection of potential recombination sites in all the six virus genera composing this family with various degrees of consistency. Furthermore, based on Fisher's Exact Test of Neutrality, no positive selection exerted on protein-coding genes was detected. Inferred phylogenies showed that putative recombination events played a major role in the promiscuous evolutionary history of the family *Alphaflexiviridae*. In fact, contrary to the statement of the International Committee on Taxonomy of Viruses, all currently the known isolates of *Potato virus X* were cleaved into two subgroups. While the smallest subgroup gathered nonrecombinant isolates, the largest encompassed recombinants. Similarly, members of *Potexvirus* genus split into three distinct groups. Although groups I and III comprised several recombinants, the group II was constituted by the two nonrecombinant isolates of *Strawberry mild yellow edge virus*. Moreover, all recombinant unclassified potexviruses were revealed to be phylogenetically closely related to recombinant typical potexviruses. As a result, a reclassification should be considered.

* Corresponding author: Moncef Boulila, Institut de l'Olivier, B.P. 14, 4061 Sousse Ibn-Khaldoun, Tunisia. phone/fax: 00216 73236135, E-mail address: boulila.moncef@ yahoo.fr.

Keywords: Bioinformatics; Phylogeny; Recombination; natural selection; *Alphaflexiviridae;* Sequence

Abbreviations

GIRE:	genomic interval of recombination events
MEGA:	Molecular Evolutionary Genetics Analysis
ME:	Minimum Evolution
MP:	Maximum Parsimony
ML:	Maximum Likelihood
NJ:	Neighbor joining
NRS:	number of recombination sites
RECCO:	Recombination Cost Optimization
UPGMA:	Unweighted Pair Group Method with Arithmetic mean

Introduction

One of the most unusual features of RNA viruses is their enormous genetic variability. Among the different processes contributing to the continuous generation of new viral variants, RNA recombination is of special importance. An increasing number of animal and plant viruses have been shown to undergo RNA-RNA recombination. The high frequency and widespread nature of RNA recombination indicate that this phenomenon plays a significant role in the biology of RNA viruses. Viral RNA recombination, a process that joins together two noncontiguous RNA segments, is an especially powerful tool in virus evolution, since it can rapidly lead to dramatic changes in virus genomes by recombining or rearranging sequences. Accordingly, the significant role of RNA recombination in emergence of new viruses or virus strains is well documented for numerous viruses of different hosts (human, animal, plant, insect, fungus, bacterium) (Ayllon et al. 1999; Desport et al. 1998; Hajjou et al. 1996; Strauss et al. 1988; Weiss et al. 1991). RNA recombination can be an efficient tool for viruses to repair viral genomes, thus contributing to virus fitness. It may also play a role in the formation of subviral RNAs that include defective interfering (DI) RNAs associated with many animal and plants viruses (Pathak and Nagy 2009) as well as with chimeric satellite RNAs (Cheng and Nagy 2003). It has been shown that RNA recombination enables the exchange of genetic material not only between the same or similar viruses but also between distinctly different viruses (Worobey and Holmes 1999). Sometimes, it also permits crossovers between viral and host RNA (Aaziz and Tepfer 1999; Baroth et al. 2000; Greene and Allison 1994; Nagai et al. 2003). Taking into account the structure of viral genomic molecules and the location of crossover sites, three basic types of RNA recombination were distinguished: homologous, aberrant homologous and non-homologous (Alejska et al. 2001; Lai 1992). The former two occur between two identical or similar RNAs (or between molecules displaying local homology), while the latter involves two different molecules. Most of the collected data suggest that RNA recombinants are formed according to a copy choice model (Alejska et al. 2001). A viral replication complex starts nascent RNA strand synthesis on one template, called RNA donor and then switches to another template, called RNA acceptor. Accordingly, two main factors are thought to affect RNA recombination: the structure of recombining molecules and the ability of the viral replicase to switch templates. RNA virus exists in a quasispecies nature, which represents microheterogeneity among viral population. The mutations occur associated with the lack of proofreading activity of the RNA polymerase during replication. The genome changes may render the viral genome able to replicate more efficiently or to escape the host immune responses (Wang et al. 2007). Positive selection usually preserves such favorable mutations. In contrast, negative selection removes deleterious substitutions from the viral populations, thereby ensuring survival of the viral species. Through generations, viral populations evolve under various selective forces at different regions and sites that display different functional constraints. A stringent and robust criterion for detecting adaptive evolution in a protein-coding gene is an accelerated nonsynonymous (d_N, amino acid replacing) rate relative to the synonymous (d_S, silent) rate of substitutions, with the rate ratio $\omega = d_N/d_S > 1$. As silent mutations do not change the amino acid whereas replacement mutations do, the difference in their fixation rates provides a measure of selective pressure on the protein. Amongst positive-strand plant RNA viruses, the family *Alphaflexiviridae* encompasses several viruses having an important economical impact. It includes the following genera: *Allexivirus*, *Botrexvirus*, *Lolavirus*, *Mandarivirus*, *Potexvirus*, and *Sclerodarnavirus*. The family *Alphaflexiviridae* includes the plant viruses with a potex-like polymerase. The chief characteristics of members of the family are: (i) flexuous filamentous not enveloped virions, 470-800 nm long and 12-13 nm in diameter, (ii) monopartite, positive sense, single-stranded RNA genomes with a 3'-poly(A) tail having a size of 5.4-9.0 kb, (iii) translation of at least some ORFs from subgenomic mRNAs (viruses in the family employ both 5'- and 3'-coterminal subgenomic mRNAs, (iv) up to six ORFs ordered from 5' to 3', (v) all viruses encode a triple gene block (TGB) set of movement proteins, (vi) a single capsid (coat) protein (CP) (22-43 kDa), (vii) a replication protein (150-196 kDa) containing conserved methyl transferase, helicase and RNA-dependent RNA polymerase (RdRp) motifs (Koonin and Dolja 1993). Biologically, the viruses are fairly diverse. They have been reported from fungi (*Botrexvirus* and *Sclerodarnavirus* genera) and from a wide range of herbaceous (members of the genera *Allexivirus*, *Lolavirus*, and *Potexvirus*) and woody plant species (*Indian citrus ringspot mandarivirus*) but the host range of individual members is usually limited. No vectors are known except for members of *Allexivirus* genus which can be transmitted by mites. Aggregates of virus particles accumulate in the cytoplasm (Adams et al. 2004; Martelli et al. 2007). The purpose of the present study was to investigate and characterize putative recombination events and selective constraints acting on the complete genome of currently the known members of the family *Alphaflexiviridae* and finally to propose a new classification based on the predicted evolutionary relationships among them. This was achieved by analyzing the genome of 73 accessions collated from GenBank.

Materials and Methods

Virus Sequence Source

The sequences of the full-length genome of a total of 73 accessions used in this study were downloaded from GenBank (Table 1).

Table 1. Current taxonomic status of the members of the family *Alphaflexiviridae* included in the study and their accession numbers

Genus (in bold) and unclassified viruses	Virus/Isolate	GenBank accession number
Allexivirus	Garlic virus A (GarV A)	NC_003375
	Garlic virus C (GarVC)	NC_003376
	Garlic virus E.YH (GarVE.YH)	NC_004012
	Garlic virus X.Korea (GarVX.Korea)	U89243
	Garlic virus X.YH (GarVX.YH)	AJ292229
	Shallot virus X (ShVX)	NC_003795
Botrexvirus	Botrytis virus X (BVX)	NC_005132
Mandarivirus	Indian citrus ringspot virus. K1 (ICRSV.K1)	NC_003093
Sclerodarnavirus	Sclerotinia sclerotiorum debilitation-associated RNA virus (SSDaRNAV)	NC_007415
Lolavirus	Lolium latent virus (LLV)	EU489641
Potexvirus	Alternanthera mosaic virus. Russia (AltMV.Russia)	FJ822136
	Alternanthera mosaic virus.SP 3-1 (AltMV.SP 3-1)	GQ179646
	Alternanthera mosaic virus. Pennsylvania (AltMV.Pennsylvania)	NC_007731
	Alternanthera mosaic virus.SP 4-7 (AltMV.SP 4-7)	GQ179647
	Cactus virus X (CVX)	NC_002815
	Cassava common mosaic virus.Brazilian (CCMV)	U23414
	Cymbidium mosaic virus.clone 18-1 (CymMV.clone 18.1)	EF125178
	Cymbidium mosaic virus.clone 18-29 (CymMV.clone 18-29).	EF125180
	Cymbidium mosaic virus clone 18-30 (CymMV.clone 18-30)	EF125179
	Cymbidium mosaic virus. Japanese (CymMV.Japanese)	AB197937
	Cymbidium mosaic virus.Korean type 2 (CymMV.Korean type 2)	AF016914
	Cymbidium mosaic virus. M2. Taiwan (CymMV.M2.Taiwan)	EU314803
	Cymbidium mosaic virus. plm1 (CymMV.plm1)	AM055720
	Cymbidium mosaic virus. Singapore (CymMV.Singapore)	U62963
	Cymbidium mosaic virus. SMi2 (CymMV.SMi2)	AM055640
	Cymbidium mosaic virus.Taiwan (CymMV.Taiwan)	AY571289
	Foxtail mosaic virus. Taiwan (FMV.Taiwan)	AY121833
	Foxtail mosaic virus (FMV)	M62730
	Foxtail mosaic virus.clone p9 (FMV.clone p9)	EF630359
	Hosta virus X. HVX-Kr (HVX-Kr)	NC_011544

Genus (in bold) and unclassified viruses	Virus/Isolate	GenBank accession number
	Hydrangea ringspot virus. PD 109 (HRSV.PD 109)	AY707100
	Lettuce virus X.Iran (LeVX.Iran)	NC_010832
	Lily virus X.Netherlands (LVX.Netherlands)	NC_007192
	Narcissus mosaic virus. New Zealand (NMV.New Zealand)	AY225449
	Narcissus mosaic virus (NMV)	NC_001441
	Nerine virus X.J (NVX.J)	NC_007679
	Phaius virus X.Japan (PhVX.Japan)	NC_010295
	Plantago asiatica mosaic virus (PlAMV)	NC_003849
	Plantago asiatica mosaic virus. Pr (PlAMV.Pr)	AB360796
	Plantago asiatica mosaic virus. Li6 (PlAMV.Li6)	AB360795
	Plantago asiatica mosaic virus. Li5 (PlAMV.Li5)	AB360794
	Plantago asiatica mosaic virus. Li4 (PlAMV.Li4)	AB360793
	Plantago asiatica mosaic virus. Li3 (PlAMV.Li3)	AB360792
	Plantago asiatica mosaic virus. Li2 (PlAMV.Li2)	AB360791
	Plantago asiatica mosaic virus. Li1 (PlAMV.Li1)	AB360790
	Nandina mosaic virus. PLH-1 (NMV.PLH-1)	AY800279
	Potato virus X.Iran (PVX.Iran)	FJ461343
	Potato virus X.Taiwan (PVX.Taiwan)	AF272736
	Potato virus X. Korean(PVX.Korean)	AF373782
	Potato virus X. CP4 (PVX.CP4)	AF172259
	Potato virus X. ROTH1 (PVX.ROTH1)	AF111193
	Potato virus X. CP (PVX.CP)	M31541
Potexvirus	Potato virus X. PVXCOPR (PVX.PVXCOPR)	M95516
	Potato virus X. South American (PVX.South American)	M63141
	Potato virus X. PVXCG (PVX.PVX.CG)	M72416
	Potato virus X. X3 (PVX.X3)	NC_011620
	Potato virus X. FX21(PVX.FX21)	EF423572
	Potato virus X. bs (PVX.bs)	AB056719
	Potato virus X. os (PVX.os)	AB056718
	Potato virus X.Tula (PVX.Tula)	EU021215
	Potato Virus X. HB (PVX.HB)	X72214
	Potato virus X. PVXGEN (PVX.PVXGEN)	M38480
	Scallion virus X.Chinese (SVX.Chinese)	NC_003400
	Strawberry mild yellow edge virus. MY-18 (SMYEV.MY-18)	NC_003794
	Strawberry mild yellow edge virus. D74 (SMYEV. D74))	AJ577359
	Tulip virus X. J (TVX)	NC_004322
Unclassified Potexvirus	Allium virus X (AlVX)	FJ670570
	Alstroemeria virus x (AVX)	NC_007408
	Asparagus virus 3.J (AV-3.J)	NC_010416
	Mint virus X (MVX)	NC_006948
	Opuntia virus X. CC10 (OVX.CC10)	NC_006060
	Schlumbergera virus X. K11 (ScVX.K11)	NC_011659
	Zygocactus virus X. B1 (ZVX.B1)	NC_006059

Sequence Alignments, Selective Constraints, Recombination and Phylogenetic Analyses

The nucleotide sequences were aligned using CLUSTALW 2.0.9 and CLUSTALX 2.0.9 (Larkin et al. 2007) with default parameters. Their phylogenetic relationships were determined by five methods using the neighbor-joining (NJ), Minimum Evolution (ME), Maximum Parsimony (MP), and Unweighted Pair Group Method with Arithmetic mean (UPGMA) implemented in MEGA4.1β program (Kumar et al. 2008) and Maximum-likelihood (ML) using PHYLO_WIN graphic tool (Galtier et al. 1996). Bootstrap analyses with 500 replicates were performed to assess the robustness of the branches.

Positive selection, using MEGA4.1β program, was inferred by the counting method described by Nei and Gojobori (1986) and later by Suzuki and Gojobori (1999). This method utilized the phylogenetic tree of sequences analyzed. In the parsimony method, the total numbers of synonymous (c_S) and nonsynonymous (c_N) substitutions as well as the average numbers of synonymous (s_S) and nonsynonymous (s_N) sites per codon over the phylogenetic tree for each codon site were computed according to the maximum parsimony principle (Fitch 1971; Hartigan 1973).

The null hypothesis of selective neutrality (r_S=r_N or $\omega = 1$) was tested for each site by computing the probability (p) of obtaining the observed or more biased values for c_S and c_N, which were assumed to follow a binomial distribution with the probabilities of occurrence of synonymous and nonsynonymous substitutions given by $s_S/(s_S + s_N)$ and $s_N/(s_S + s_N)$, respectively. Positive selection is inferred when $p<0.05$ and $c_N/s_N>c_S/s_S$ (Suzuki 2006).

Potential recombination events between diverged nucleotide sequences were explored with two programs: RDP version 3.31β (Martin et al. 2005b) and RECCO (Maydt and Lengauer 2006). RDP incorporates several published recombination detection methods into a single suite of tools: RDP (Martin and Rybicki 2000), GENECONV (Padidam et al. 1999), BOOTSCAN (Martin et al. 2005a), MAXCHI (Smith 1992), CHIMAERA (Posada and Crandall 2001), SISCAN (Gibbs et al. 2000), and 3SEQ (Boni et al. 2007). In all cases, defaults parameters were used. Only events predicted by more than half of the methods are considered as significant. The algorithm developed and described by Maydt and Lengauer (2006) as fast, simple and sensitive method for detecting recombination in a set of sequences and locating putative recombination breakpoints is based on cost minimization. This method has only two tunable parameters, recombination and mutation cost. In practice the only parameter considered is α, representing the cost of mutation relative to recombination. When α changes from 0 to 1, the cost of mutation weighted by α increases, and the cost for recombination weighted by 1- α decreases. In other words the parameter α controls the ambiguity between mutation and recombination.

Table 2. Determination of inferred putative recombination events and their frequency along the sequences of the entire genome of members of the family *Alphaflexiviridae*

Recombination determined by RECCO								Recombination determined by RDP v3.31β
Virus.isolate			Length of breakpoint				Genomic position of the longest breakpoint (size in nucleotide)	putative parentals (Major x Minor)
	NRS	GIRE (nt)	1 residue	2 residues	3 residues	>3residues		
GarVA	57	907-10111	19	21	6	11	10034-10045(12)	/
GarVC	65	1051-10467	17	5	15	28	10424-10467(43)	/
GarVE.YH	61	1051-10439	16	11	11	23	6502-6521(20)	GarVX.YH x GarV A
GarVX.Korea	2	8057 8261	2	0	0	0	8057(1) 8261(1)	/
GarVX.YH	5	8057-8263	1	2	0	2	8217-8225(9)	/
ShVX	52	903-10340	18	12	6	16	9845-9853(9)	/
BVX	273	955-9441	83	57	47	86	3401-3427(27)	/
ICRSV.K1	192	917-9589	75	37	23	57	3938-3966(29)	/
SSDaRNAV	175	948-8055	56	35	26	58	2673-2762(90)	/
AltMV.Russia	1	4927-4953	0	0	0	1	4927-4953(27)	/
AltMV.SP 3-1	1	4741-4812	0	0	0	1	4741-4812(72)	AltMV.Russia x AltMV.SP 4-7
AltMV. Pennsylvania	2	2331-2892	1	0	0	1	2874-2892(19)	/
AltMV.SP 4-7	1	4747-4757	0	0	0	1	4747-4757(11)	/
CVX	21	932-9028	7	4	3	7	3394-3433(40)	/
CCMV	124	1341-8764	40	26	18	40	6691-6822(132)	/
CymMV.clone 18.1	2	1415-8321	0	0	0	2	1415-1439(25)	CymMV.clone 18-30 x CymMV.Singapore ICRSV.K1 x CVX
CymMV.clone 18-29	10	3605-8386	0	0	0	10	3605-3879(274)	ICRSV.K1 x CVX
CymMV.clone 18-30	2	8434-8958	0	0	0	2	8769-8958(190)	ICRSV.K1 x CVX
CymMV. Japanese	2	3336-3879	0	0	0	2	3531-3879(348)	ICRSV.K1 x CVX

Table 2. (Continued)

Recombination determined by RECCO								Recombination determined by RDP v3.31β
Virus.isolate			Length of breakpoint				Genomic position of the longest breakpoint (size in nucleotide)	putative parentals (Major x Minor)
	NRS	GIRE (nt)	1 residue	2 residues	3 residues	>3residues		
CymMV.Korean type 2	1	6601-6602	0	1	0	0	6601-6602(1)	ICRSV.K1 x CVX
CymMV.M2. Taiwan	1	3081-3161	0	0	0	1	3081-3161(81)	ICRSV.K1 x CVX
CymMV.plm1	9	4425-8433	0	0	0	9	4498-4713(216)	ICRSV.K1 x CVX
CymMV. Singapore	8	1926-8351	0	0	0	8	6430-6600	ICRSV.K1 x CVX
CymMV.SMi2	5	3363-6600	0	0	0	5	6474-6600(127)	ICRSV.K1 x CVX
CymMV.Taiwan	6	4219-7342	0	0	0	2	4498-4551(54)	ICRSV.K1 x CVX CymMV.clone 18-29 x CymMV.clone 18-30
FMV.Taiwan	2	8184-8594	0	0	0	2	8184-8392(209)	/
FMV	3	7847-8403	2	0	1	0	7847-7849(3)	/
FMV.clone p9	2	8378-9037	1	0	0	1	8378-8392(5)	/
HVX	187	1024-8993	61	38	31	57	3758-3813(56)	/
HRSV	128	1006-9016	34	22	17	55	6487-6645(159)	/
LeVX.Iran	37	912-8946	10	6	5	16	4741-4756(16)	ICRSV.K1 x CVX
LVX.Netherlands	81	1192-9005	23	25	12	21	1646-1664(19)	/
NVX.J	288	1051-8968	112	48	46	82	6692-6768(77)	/
PhVX.Japan	66	1192-8896	15	11	17	23	4494-4506(13)	/
PlAMV	30	1117-8780	2	4	3	21	3113-3169(57)	/
NMV.PLH-1	26	1117-8968	7	2	3	13	3688-3752(65)	/
PVX.Iran	15	2089-8543	0	0	3	12	7122-7193(72)	/
PVX.Korean	3	2807-4442	0	0	2	1	4435-4442(8)	/
PVX.CP4	5	8274-8406	2	1	1	1	8274-8280(7)	/
PVX.ROTH1	11	1909-8628	0	0	2	9	8562-8628(67)	/

Recombination determined by RECCO								Recombination determined by RDP v3.31β
Virus.isolate			Length of breakpoint				Genomic position of the longest breakpoint (size in nucleotide)	putative parentals (Major x Minor)
	NRS	GIRE (nt)	1 residue	2 residues	3 residues	>3residues		
PVX.PVX.CG	13	8988-9871	0	4	1	8	9860-9871(12)	/
PVX.X3	1	1085-1090	0	0	0	1	1085-1090(6)	/
PVX.FX21	4	3525-7713	0	0	0	4	7627-7713(87)	/
PVX.bs	3	8487-8988	0	0	0	3	8804-8988(185)	/
PVX.os	2	8589-8988	0	0	0	2	8741-8988(248)	/
PVX.Tula	5	2202-7160	0	1	0	4	2202-2263(60)	/
PVX.PVXGEN	1	8707-8988	0	0	0	1	8707-8988(82)	/
SVX.Chinese	12	3684-8136	1	2	5	4	8121-8132(12)	ICRSV.K1 x CVX
TVX	35	1158-8874	7	2	10	16	4255-4291(37)	/
AlVX	247	926-8966	79	50	36	82	5710-5743(34)	/
AVX	83	2399-8901	26	19	13	25	7550-7566(17)	ICRSV.K1 x CVX
AV-3.J	13	3333-9005	5	0	4	4	6380-6395(16)	ICRSV.K1 x CVX
MVX	114	957-8565	42	15	16	41	7392-7401(10)	/
OVX.CC10	63	904-8740	13	7	15	28	6138-6159(22)	/
ScVX.K11	57	1191-9051	12	8	9	28	3394-3430(37)	/
ZVX.B1	18	5005-8868	5	2	2	9	5712-5745(34)	/
LLV	269	920-9046	94	58	35	82	5720-5742(23)	/

With RDP v3.31β algorithm, only events supported by more than half of the different methods are reported. Nucleotide numbering corresponds to the aligned sequences.

Abbreviations: NRS: number of recombination sites, GIRE: genomic interval of recombination events (the span of sequences in the genome of the virus where recombination events were predicted).

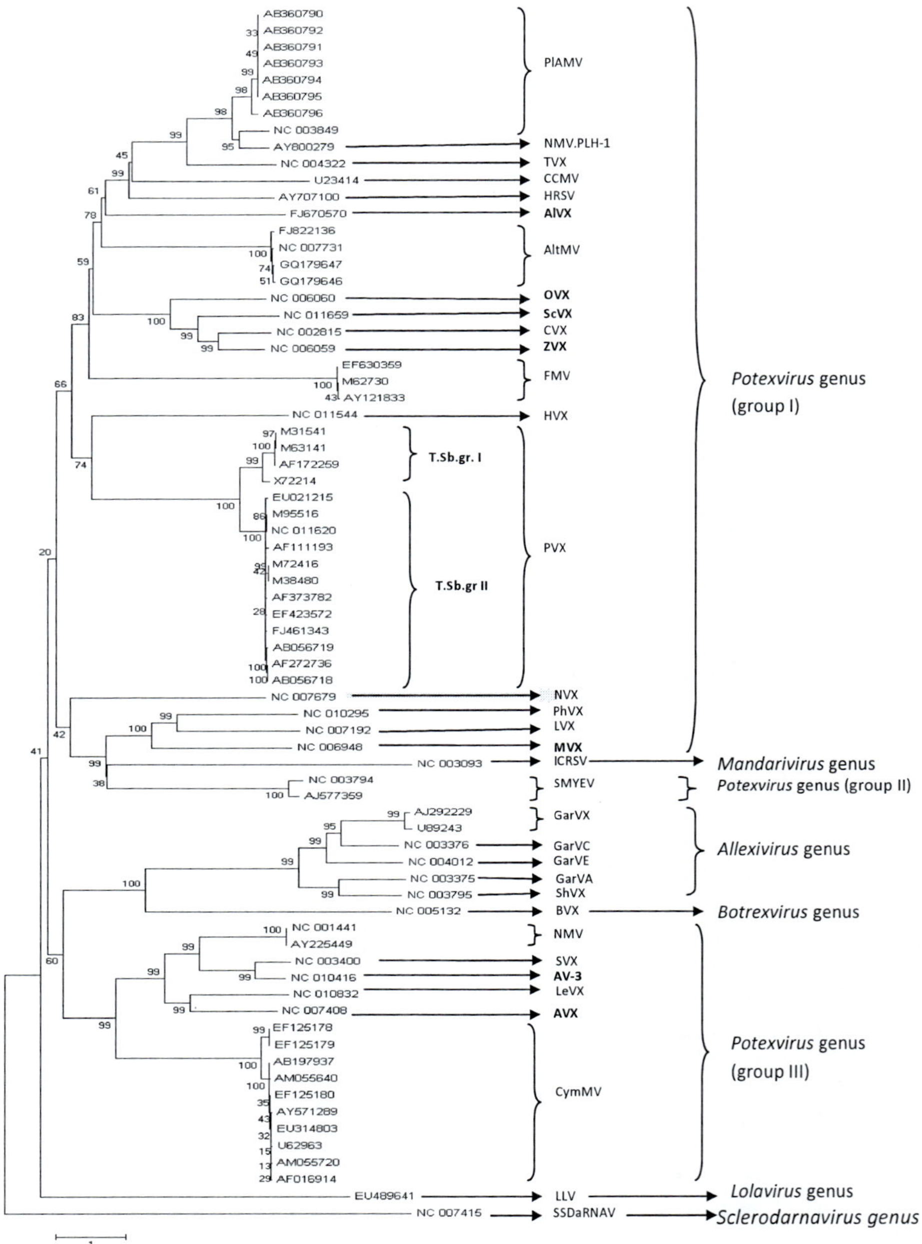

Figure 1. Dendrogram depicting phylogenetic relationships among the studied members of the family *Alphaflexiviridae* based on the analysis of their complete genome sequences. Members of *Allexivirus*, *Botrexvirus*, *Mandarivirus*, *Sclerodarnavirus*, and *Lolavirus* genera fell into five clearly delineated clusters. Nevertheless, the genus *Potexvirus* cleaved into three groups (groups I, II, III). In additions, the species *Potato virus X* was shown to be constituted by two subclusters Tentative Subgroup I (T.Sb.gr. I) and Tentative Subgroup II (T. Sb.gr. II). The tree was produced using the NJ algorithm option of MEGA4.1β. The numbers above the branches indicate the bootstrap confidence value. The scale bar shows the number of substitutions per nucleotide.

Results

Recombination Events during Evolution of the *Alphaflexiviridae*

Examination of the RECCO program output regarding the occurrence of recombination events in the entire genome of the *Alphaflexiviridae* revealed that all representatives of the genera *Allexivirus*, *Botrexvirus*, *Mandarivirus*, *Sclerodarnavirus*, *Lolavirus* as well as Unclassified *Potexvirus* members, were putative recombinants. RDP package confirmed these results only for the following accessions: GarVE.YH, AVX, and AV-3.J (Table 2). Regarding *Potexvirus* genus, 70% (39/56) of assessed accessions were potential recombinants. RDP v3.31 algorithm confirmed these results only in 33% (13/39) of putative recombining accessions. It also pointed out that ICRSV and CVX were major possible ancestors of CymMV, LeVX, SVX as well as of the unclassified poteviruses AVX and AV-3. It is worth noting that members of the genera *Allexivirus*, *Botrexvirus*, *Mandarivirus*, *Sclerodarnavirus*, *Lolavirus* along with Unclassified *Potexvirus*, possessed very frequent breakpoints which in majority consisted of a single and more than three nucleotides. Conversely, putative recombination events were scarce in most of analyzed potexviruses particularly in AltMV, CymMV, FMV, SVX, and PVX. Most of breakpoints consisted of more than three nucleotides.

Nucleotide Sequence Analysis

Maximum composite likelihood estimate of the pattern of nucleotide substitution were conducted in MEGA4.1β. The results for the *Alphaflexiviridae* were as follows: rates of different transitional substitutions varied from 4.38 to 47.85, and those of transversional substitutions varied from 0.57 to 0.76. The nucleotide frequencies were as follows: 0.280 (A), 0.220 (T/U), 0.286 (C), and 0.214 (G). The transition/transversion rate ratios were $k_1 = 7.689$ (purines) and $k_2 = 62.725$ (pyrimidines). The overall transition/transversion bias was $R = 18.477$, where $R = [AGk_1 + TCk_2]/[(A+G)(T+C)]$. There were a total of 3942 positions in the final dataset. In all these analyses, codon positions included were first + second + third + noncoding. All positions containing gaps and missing data were eliminated from the dataset (complete deletion option).

The MEGA4.1β program incorporates also the Tajima's Neutrality Test. The purpose of this test is to indentify sequences which do not fit the neutral theory model at equilibrium between mutation and genetic drift. Tajima's test compares a standardized measure of the total number of segregating sites (these are the DNA sites that are polymorphic) in the sampled DNA and the average number of mutations between pairs in the sample. The Tajima's D was determined ($D = 4.867055$).

Positive Selection

The high genetic stability of viruses can be attributed to negative or purifying selection to maintain the functional integrity of the viral genome. The degree of negative selection in genes, or the degree of functional constraint for the maintenance of the encoded protein sequence, can be estimated, as indicated above, by the ratio between the nucleotide diversities in nonsynonymous and synonymous positions (d_N/d_S). For most coding genes the d_N/d_S ratio is below unity which is consistent with negative selection against protein change. In contrast, a d_N/d_S ratio above unity may be an indication that adaptive or positive selection is driving gene divergence. In this study, pairwise comparisons of all screened accessions showed that none of members of the family *Alphaflexiviridae* was under positive selection based on results obtained by testing the neutrality in sequence pairs with Fisher's Exact Test. The probability of rejecting the null hypothesis of strict-neutrality ($d_N = d_S$) in favor of positive selection for each sequence pair was determined. Values of P less than 0.05 were considered significant at the 5% level. The variance of the difference (d_N-d_S) was computed using the bootstrap method (500 replicates). All analyses were conducted using the Nei-Gojobori method incorporated in MEGA program. All positions containing gaps and missing data were eliminated from the dataset (complete deletion option). There were a total of 900 codons in the final dataset.

Phylogenetic Relationships

The phylogenetic relationships among members of the family *Alphaflexiviridae*, based on the sequences of their complete genome, were inferred using NJ, ME, MP, ML, and UPGMA. The constructed trees basically gave a somewhat similar phylogenetic grouping for the five methods used in this study. As shown in Figure 1, only members of the *Potexvirus* genus diverged and segregated into three different subgroups to which the names group I, II and III were assigned. Interestingly, while groups I and II encompassed several putative recombinants, the two analyzed accessions of NMV were no potentially recombining virus isolates. Similarly, within the group I, the isolates of PVX retrieved into two subclusters. Although the smallest clade (Tentative Subgroup I) was composed by possible nonrecombinants, the largest clade (Tentative Subgroup II) comprised recombinant components. Furthermore, by contrast to the current classification provided so far by the International Committee on Taxonomy of Viruses (ICTV), members of Unclassified Potexvirus group dispatched in groups I and II of the same genus. Whereas AlVX, OVX, ScVX, ZVX, and MVX belonged to the group I, AV-3, and AVX incorporated into the group II. It is worth pointing out that, being themselves putative recombinants as determined by RECCO method, all unclassified potexviruses were phylogenetically closely related to recombining typical potexviruses.

Discussion

Viruses represent the epitome of genetic diversity encompassing both DNA and RNA genomes. Their variability probably results from high mutation rates, extensive genetic recombination, gene duplication, large population and ancient origin. In addition, at least a third of all viral genes may have *de novo* origins. There is also recent evidence that additional novelty may be generated by genetic recombination between viruses and their plant hosts (Mayo and Jolly 1991). Viruses are indeed master explorers of evolutionary space.

In order to enhance the knowledge on evolution of currently the known members of the family *Alphaflexiviridae*, in-depth molecular investigations in their entire genome were conducted. The molecular evolution of viruses is driven by two major forces: recombination and positive selection. An essential step in any phylogeny-based analysis is to screen for and quantify evidence for recombination (Kosakovsky Pond et al. 2006). In recent years, there has been an increased interest in understanding the role of recombination in the evolution of field populations of RNA plant viruses. Recombination may result in the exchange of long nucleotide sequences and it could have bigger phenotypic effects than most mutations. This could jeopardize the efficiency of current control strategies, particularly so for the use of resistance to viruses bred in crop varieties (Garcia-Arenal and McDonald 2003). In this study, two methods were chosen (RECCO and RDP v.3.31β) based on the fact that they are appropriate for the mosaic structure of viruses as reported in previous works (Boulila 2009a,b,c; 2010). Thus, the recombination frequency along the sequences of screened accessions of the *Alphaflexiviridae* was dissected and basically revealed that: (i) putative recombination events were detected in the genome of all members of the genera *Allexivirus*, *Botrexvirus*, *Mandarivirus*, *Sclerodarnavirus*, *Lolavirus* along with the Unclassified *Potexvirus* group, whereas only 70% of the typical potexviruses were potential recombinants, (ii) the former five virus genera together with the unclassified potexviruses were highly recombinant contrary to most of assessed recombinant typical potexviruses, (iii) exchange of heredity material might be occurring between two viruses belonging to two different virus genera. For instance, it was shown that ICRSV and CVX were possible ancestors of CymMV, LeVX, SVX as well as of the unclassified poteviruses AVX and AV-3.

The ascertaining of the selection pressures that have shaped genetic variation forms a major part of many studies of molecular evolution. A common approach to this problem involves estimating the rates of nonsynonymous (d_N) and synonymous (d_S) substitutions. Kimura (1977) was the first to emphasize that their comparison can be a powerful tool for studying natural selection. Estimates of d_N significantly different from d_S, provide convincing evidence for nonneutral evolution. This approach is attractive as it does not make any assumptions regarding the demographic history of the population, unlike many neutrality tests [e.g. Tajima's test (1989)] which compare estimates of effective population size obtained using different measures of genetic variation. In this study and according to Fisher's Exact Test of Neutrality, no positive selection operating on protein expression of all screened virus genomes was identified.

Both recombination and selective constraints can influence the evolutionary relationships among different viruses. Revealing the evolutionary history of the members of RNA virus families has become one of the main goals for evolutionary virologists (Costa-Mattioli et al. 2002; Katayama et al. 2002; Lukashov and Goudsmit 2002; Thiery et al. 2004; de Villiers et al. 2004), and the proposals for classification have been changing constantly in recent years. Many causes of this change are: (i) the lack of complete genome sequences in databanks leading to focusing on the analysis of a single gene, (ii) the presence of extensive recombination, (iii) the poor reliability of alignments due to the typically high sequence divergence of viruses and the existence of deletions and insertions. However, new techniques have been recently developed for analyzing unaligned multiple whole genomes and for producing global phylogenies (Qi et al. 2004a,b; Stuart et al. 2002a,b; Stuart and Berry 2003).

In this study, it was demonstrated that phylogenetic relationships and clustering were mainly orientated by putative recombination events that arose in the complete genome of numerous viruses such as PVX, and all unclassified potexviruses. Consequently, this is call for a possible reclassification by the ICTV.

Finally, this article provided a detailed study on molecular evolution of currently the known filamentous plant RNA viruses of the family *Alphaflexiviridae*. This was achieved by: (i) identifying putative recombination events in several accessions extracted from GenBank, (ii) dissecting their frequency along the sequences, (iii) determining the possible ancestors involved in their genetic diversity, (iv) detecting the selective pressures exerted on protein expression, (v) analyzing their evolutionary relationships.

References

Aaziz R, Tepfer M (1999) Recombination in RNA viruses and in virus-resistant transgenic plants. *J. Gen. Virol.* 80: 1339-1346.

Adams MJ, Antoniw JF, Bar-Joseph M, Brunt AA, Candresse T, Foster GD, Martelli GP, Milne RG, Zavriev SK, Fauquet CM (2004) The new plant virus family *Flexiviridae* and assessment of molecular criteria for species demarcation. *Arch. Virol.* 149: 1045-1060.

Alejska M, Kurzyniska-Kokorniak A, Broda M, Kierzek R, Figlerowicz M (2001) How RNA viruses exchange their genetic material. *Acta Bioch. Polon.* 48: 391-407.

Ayllon MA, Lopez C, Navas-Castillo J, Mawassi M, Dawson DO, Guerri J, Flores R, Moreno P (1999) New defective RNAs from *Citrus tristeza virus*: evidence for a replicase driven template switching mechanism in their generation. *J. Gen. Virol.* 80: 817-821.

Baroth M, Orlich M, Thiel HJ, Becher P (2000) Insertion of cellular NEDD8 coding sequences in a pestivirus. *Virology* 278: 456-466.

Boni MF, Posada D, Feldman MW (2007) An exact nonparametric method for inferring mosaic structure in sequence triplets. *Genetics* 176: 1035-1047.

Boulila M. (2009a) Molecular characterization of an almond isolate of *Prune dwarf virus* in Tunisia: putative recombination breakpoints in the partial sequences of the coat protein-encoding gene in isolates from different geographic origin. *Phytopathol. Mediterr.* 48: 411-421.

Boulila M. (2009b) Molecular evidence for recombination in *Prunus necrotic ringspot virus*. *Plant Mol. Biol. Rep.* 27: 189-198.

Boulila M. (2009c) Recombination structure and genetic relatedness among members of the family *Bromoviridae* based on their RNAs 1 and 2 sequence analyses. *Virus Genes* 38(3): 435-444.

Boulila M. (2010) Putative recombination events and evolutionary history of five economically important viruses of fruit trees based on the coat protein-encoding gene sequence analysis. *Biochem. Genet.* 48: 357-375.

Cheng C P, Nagy PD (2003) Mechanism of RNA recombination in Carmo- and Tombusviruses: evidence for template switching by the RNA-dependant RNA polymerase in vitro. *J. Virol.* 77(22): 12033-12047.

Costa-Mattioli M, Cristina J, Romero H, Perez-Bercof R, Casane D, Colina R, Garcia L, Vega I, Glikman G, Romanowsky V, Castello A, Nicand E, Gassin M, Billaudel S, Ferre

V (2002) Molecular Evolution of *Hepatitis A virus*: a new classification based on the complete VP1 protein. *J. Virol.* 76: 9516-9525.

Desport M, Collins ME, Brownlie J (1998) Genomic instability in BVDV: an examination of the sequence and structural influences on RNA recombination. *Virology* 246: 352-361.

de Villiers EM, Fauquet C, Broker TR, Bernard HU, zur Hausen H (2004) Classification of papillomaviruses. *Virology* 324: 17-27.

Fitch, W.N. (1971). Toward defining the course of evolution minimum change for a specific tree topology. *Syst. Zool.* 20: 406-416.

Galtier N, Gouy M, Gautier C (1996) SEAVIEW and PHYLO_WIN: two graphic tools for sequence alignment and molecular phylogeny. *CABIOS* 12(6): 543-548.

Garcia-Arenal F, McDonald BA (2003) An analysis of the durability of the resistance to plant viruses. *Phytopathology* 93: 941-952.

Gibbs MJ, Armstrong JS, Gibbs AJ (2000) Sister-scanning: a Monte Carlo procedure for assessing signals in recombinant sequences. *Bioinformatics* 16: 573-582.

Greene AE, Allison RF (1994) Recombination between viral RNA and transgenic plants transcripts. *Science* 263: 1423-1425.

Hajjou M, Hill JR, Subramaniam SV, Hu JY, Raju R (1996) Non homologous RNA-RNA recombination events at the 3' nontranslated region of the sindbis virus genome: hotspots and utilization of nonviral sequences. *J. Virol.* 70: 5153-5164.

Hartigan JA (1973) Minimum mutation fits to a given tree. *Biometrics* 29: 53-63.

Katayama K, Shirato-Horikochi H, Kojima S, kageyama T, Oka T, Hoshino FB, Fukushi S, Shinohara M, Uchida K, Suzuki Y, Gojobori T, Takeda N (2002) Phylogenetic analysis of the complete genome of 18 Norwalk-like viruses. *Virology* 299: 225-239.

Kimura M (1977) Preponderance of synonymous changes as evidence for the neutral theory of molecular evolution. *Nature* 38: 515-524.

Koonin EV, Dolja VV (1993) Evolution and taxonomy of positive-strand RNA viruses: implication of comparative analysis of amino acid sequences. *Crit. Rev. Biochem. Mol. Biol.* 28: 375-430.

Kosakovsky Pond SL, Posada D, Gravenor MB, Woelk CH, Frost SDW (2006) Automated phylogenetic detection of recombination using a genetic algorithm. *Mol. Biol. Evol.* 23(10): 1891-1901.

Kumar S, Nei M, Dudley J, Tamura K (2008) MEGA: A biologist-centric software for evolutionary analysis of DNA and protein sequences. *Brief Bioinfor.* 9(4): 299-306.

Lai MMC (1992) RNA recombination in animal and plant viruses. *Microbiol. Rev.* 56: 61-79.

Larkin MA, Blackshileds G, Brown NP, Chenna R, McGettigan PA, McWilliam H, Valentin F, Wallace IM, Wilm A, Lopez R, Thompson JD, Gibson TJ, Higgins DG (2007) Clustal W and Clustal X version 2.0. *Bioinformatics* 23(21): 2947-2948.

Lukashov VV, Goudsmit J (2002) Evolutionary relationships among *Astriviridae*. *J. Gen. Virol.* 83: 1397-1405.

Martin DP, Posada D, Crandall KA, Williamson C (2005a) A modified bootscan algorithm for automated identification of recombination sequences and recombination breakpoints. *AIDS Res. H Retrovir.* 21: 98-102.

Martin DP, Williamson C, Posada D (2005b) RDP2: recombination detection and analysis from sequence alignments. *Bioinformatics* 21: 260-262.

Mayo MA, Jolly CA (1991) The 5'-terminal sequence of *Potato leafroll virus* RNA: evidence of recombination between virus and host RNA. *J. Gen. Virol.* 72: 2591-2595.

Maydt J, Lengauer T (2006) Recco: recombination analysis using cost optimization. *Bioinformatics* 22(9): 1064-1071.

Nagai M, Sakoda Y, Mori M, Hayashi M, Kida H, Akashi H (2003) Insertion of a cellular sequence and RNA recombination in the structural protein coding region of cytopathogenic bovine viral diarrhea virus. *J. Gen. Virol.* 84Pt2: 447-452.

Nei M, Gojobori T (1986) Simple methods for estimating the numbers of synonymous and nonsynonymous nucleotide substitutions. *Mol. Biol. Evol.* 3(5): 418-426.

Padidam M, Sawyer S, Fauquet CM (1999) Possible emergence of new geminiviruses by frequent recombination. *Virology* 265: 218-225.

Pathak KB, Nagy PD (2009) Defective interfering RNAs: foes of viruses and friends of virologists. *Viruses* 1: 895-919.

Posada D, Crandall K (2001) Evaluation of methods for detecting recombination from DNA sequences. Computer simulation. *Proc. Natl. Acad. Sci. USA* 98: 13757-13762.

Qi J, Luo H, Hao B (2004a) CVTree: a phylogenetic tree reconstruction tool based on whole genomes. *Nucleic Acids Res.* 32: W45-W47.

Qi J, Wang B, Hao B (2004b) Whole genome prokaryote phylogeny without sequence alignment: a K-string composition approach. *J. Mol. Evol.* 58: 1-11.

Smith JM (1992) Analyzing the mosaic structure of genes. *J. Mol. Evol.* 34: 126-129.

Strauss JH, Strauss EG (1988) Evolution of RNA viruses. *Ann. Rev. Microbiol.* 42: 657-683.

Stuart GW, Moffet K, Baker S (2002a) Integrated gene and species phylogenies from unaligned whole genome protein sequences. *Bioinformatics* 18: 100-108.

Stuart GW, Moffet K, Leader JJ (2002b) A comprehensive vertebrate phylogeny using vector representations of proteins sequences from whole genome. *Mol. Biol. Evol.* 19: 554-562.

Stuart GW, Berry MW (2003) A comprehensive whole genome bacterial phylogeny using correlated peptide motifs defined in a high dimensional vector space. *J. Bioinfor. Comp. Biol.* 1: 475-493.

Suzuki Y (2006). Statistical properties of the methods for detecting positively selected amino acid sites. *Gene* 365: 125-129.

Suzuki Y, Gojobori T (1999) A method for detecting positive selection at single amino acid sites. *Mol. Biol. Evol.* 16(10): 1315-1328.

Tajima F (1989) Statistical-method for testing the neutral mutation hypothesis by DNA polymorphism. *Genetics* 123: 585-595.

Thiéry R, Cozien J, de Boisséson C, Kerbart-Bosher S, Névarez L (2004) Genomic classification of new betanodavirus isolates by phylogenetic analysis of the coat protein suggests a low host-fish species specificity. *J. Gen. Virol.* 85: 3079-3087.

Wang SY, Wu JC, Chiang TY, Huang YH, Su CW, Sheen IJ (2007) Positive selection of Hepatitis delta antigen in chronic Hepatitis D patients. *J. Virol.* 81(9): 4438-4444.

Weiss BG, Schlesinger S (1991) Recombination between Sindbis virus RNAs. *J. Virol.* 65: 4017-4025.

Worobey M, Holmes EC (1999) Evolutionary aspects of recombination in RNA viruses. *J. Gen. Virol.* 80: 2535-2543.

In: Bioinformatics Research: New Developments
Editors: C. Battik and K. Belhassine
ISBN: 978-1-61942-363-3

Chapter VI

Molecular Evolution and Genetic Distinctness among Rod-shaped Plant RNA Viruses of the Family *Virgaviridae*

Moncef Boulila*
Institut de l'Olivier B.P. 14; 4061 Sousse Ibn-khaldoun, Tunisia

Abstract

In assessing putative recombination events and their frequencies in the complete genome of 168 accessions extracted from the international databases, in-depth molecular investigations were carried out by using two algorithms adapted to the mosaic structure of viruses: RECCO and RDP version 3.31β. By dissecting the recombination frequency through the sequences, it was demonstrated that the six genera of the family *Virgaviridae* encompass numerous members subjected to recombination. It was revealed that: (i) recombination events were widespread in the genome of several members of the six genera composing this family, (ii) recombination events were more frequent in members of *Furovirus*, *Pecluvirus*, *Pomovirus*, and *Tobravirus* genera than those of *Hordeivirus* and *Tobravirus* genera, (iii) in majority, swapped fragments had a length exceeding three nucleotides reaching a size of as long as 952 residues (*Tobacco mosaic virus*.152 isolate), (iv) as assessed by RDP package, heredity material exchange arose between non homologous sequences like those of *Indian peanut clump* and *Peanut clump* *pecluvirues.* Furthermore, based on Fisher's Exact test of Neutrality, positive selection operating of protein expression was detected only in *Beet soil-borne* and *Potato mop-top pomoviruses* as well as in the typical tobamovirus *Tomato mosaic virus* and the unclassified tobamovirus *Maracuja mosaic virus*. According to intra-genera analyses, inferred phylogenies showed that: : (i) within segmented viruses, each delineated cluster corresponded to a divided genomic region, (ii) *Tobacco mosaic virus* isolates retrieved into three groupings, (iii) *Tomato mosaic* and *Tobacco mosaic tobamoviruses* may have

* Corresponding author: Moncef Boulila, Institut de l'Olivier, B.P. 14, 4061 Sousse Ibn-Khaldoun, Tunisia. phone/fax: 00216 73236135, E-mail address: boulila.moncef@ yahoo.fr.

common ancestors, (iv) unclassified tombusvirus members did not constitute a phylogenetically homogenous group.

Keywords: Bioinformatics; Phylogeny; Recombination; natural selection; *Virgaviridae* Sequence

Introduction

The severe economic consequences of emerging plant viruses highlight the importance of studies of plant virus evolution. One question of particular relevance is the extent to which the genomes of plant viruses are shaped by recombination. Viral RNA recombination, a process that joins together two noncontiguous RNA segments, is an especially powerful tool in virus evolution, since it can rapidly lead to dramatic changes in virus genomes by recombining or rearranging sequences. Accordingly, the significant role of RNA recombination in emergence of new viruses or virus strains is well documented for numerous viruses of different hosts (human, animal, plant, insect, fungus, bacterium) (Strauss et al., 1988; Weiss et al., 1991; Hajjou et al., 1996; Desport et al., 1998; Ayllon et al., 1999). RNA recombination can be an efficient tool for viruses to repair viral genomes, thus contributing to virus fitness. It may also play a role in the formation of subviral RNAs that include defective interfering (DI) RNAs associated with many animal and plants viruses (Pathak and Nagy 2009) as well as with chimeric satellite RNAs (Cheng and Nagy 2003). It has been shown that RNA recombination enables the exchange of genetic material not only between the same or similar viruses but also between distinctly different viruses (Worobey and Holmes 1999). Sometimes, it also permits crossovers between viral and host RNA (Greene and Allison 1994; Aaziz and Tepfer 1999, Baroth et al., 2000, Nagai et al., 2003). Taking into account the structure of viral genomic molecules and the location of crossover sites, three basic types of RNA recombination were distinguished: homologous, aberrant homologous and non-homologous (Lai 1992; Alejska et al. 2001). The former two occur between two identical or similar RNAs (or between molecules displaying local homology), while the latter involves two different molecules. Most of the collected data suggest that RNA recombinants are formed according to a copy choice model (Alejska et al. 2001). A viral replication complex starts nascent RNA strand synthesis on one template, called RNA donor and then switches to another template, called RNA acceptor. Accordingly, two main factors are thought to affect RNA recombination: the structure of recombining molecules and the ability of the viral replicase to switch templates.

RNA virus exists in a quasispecies nature, which represents microheterogeneity among viral population. The mutations occur associated with the lack of proofreading activity of the RNA polymerase during replication. The genome changes may render the viral genome able to replicate more efficiently or to escape the host immune responses (Wang et al. 2007). Positive selection usually preserves such favorable mutations. In contrast, negative selection removes deleterious substitutions from the viral populations, thereby ensuring survival of the viral species. Through generations, viral populations evolve under various selective forces at different regions and sites that display different functional constraints. A stringent and robust criterion for detecting adaptive evolution in a protein-coding gene is an accelerated nonsynonymous (d_N, amino acid replacing) rate relative to the synonymous (d_S, silent) rate of substitutions, with the rate ratio $\omega = d_N/d_S > 1$. As silent mutations do not change the amino acid whereas replacement mutations do, the difference in their fixation rates provides a measure of selective pressure on the protein.

Amongst positive-strand plant RNA viruses, the family *Virgaviridae* encompasses several viruses having an important economical impact. It includes the following genera: *Furovirus*, *Hordeivirus*, *Pecluvirus*, *Pomovirus*, *Tobamovirus*, and *Tobravirus*. The chief characteristics of members of the family are: (i) alpha-like replication proteins that form a distinct phylogenetic family (Koonin and Dolja, 1993), (ii) single-stranded positive sense RNA genome with a 3'-t-RNA-like structure and no poly A tail, (iii) rod-shaped virions 20-25 nm in diameter with a central "canal", (iv) coat proteins of 19-24 kDa. The family *Virgaviridae* contains some viruses in which there is a single cell-to-cell movement protein (MP) of the '30K' superfamily (*Furovirus*, *Tobamovirus*, and *Tobravirus* genera) (Melcher, 2000) and others that encode a triple gene block (TGB) (*Hordeivirus*, *Pecluvirus*, and *Pomovirus* genera) (Morozov and Solovyer, 2003). There are also differences in the number of genomic RNAs: one (*Tobamovirus*), two (*Furovirus*, *Pecluvirus*, and *Tobravirus*), and three (*Hordeivirus* and *Pomovirus*). Biologically, the viruses are fairly diverse. They have been reported from a wide range of herbaceous and mono- and dicotyledonous plant species, but the host range of individual members is usually limited. All members can be transmitted experimentally by mechanical inoculation, and for those in the genus *Tobamovirus*, this is the only known means of transmission. In some genera, transmission is by soil-borne vectors, while members of the genus *Hordeivirus* are transmitted through pollen and seed (Adams et al., 2009).

The purpose of this study was to shed lights on molecular evolution of members of the family *Virgaviridae* dealing with selective pressure and putative recombination events by dissecting their frequencies in the genome of 168 accessions retrieved from the international databases and finally to determine their phylogenetic reconstruction.

Materials and Methods

Virus Sequence Source

The sequences of the full-length genome of a total of 168 accessions used in this study were downloaded from GenBank (Table 1).

Sequence alignments, selective constraints, recombination and phylogenetic analyses.

The nucleotide sequences were aligned using CLUSTALW 2.0.9 and CLUSTALX 2.0.9 (Larkin et al. 2007) with default parameters. Their phylogenetic relationships were determined by five methods using the neighbor-joining (NJ), Minimum Evolution (ME), Maximum Parsimony (MP), and Unweighted Pair Group Method with Arithmetic mean (UPGMA) implemented in MEGA4.1β program (Kumar et al. 2008) and Maximum-likelihood (ML) using PHYLO_WIN graphic tool (Galtier et al. 1996). Bootstrap analyses with 500 replicates were performed to assess the robustness of the branches.

Positive selection, using MEGA4.1β program, was inferred by the counting method described by Nei and Gojobori (1986) and later by Suzuki and Gojobori (1999). This method utilized the phylogenetic tree of sequences analyzed.

Table 1. Current taxonomic status of members of the family *Virgaviridae* included in the study and their accession numbers

Genus (in bold) and unclassified viruses	Virus.Isolate	GenBank accession number
Furovirus	Chinese wheat mosaic virus RNA 1/Yantai-China (CWMV RNA 1/Yantai)	NC_002359
	Chinese wheat mosaic virus RNA 2/Yantai-China (CWMV RNA 2/Yantai)	NC_002356
	Chinese wheat mosaic virus RNA1 /Rongcheng (CWMV RNA 1/ Rongcheng)	AJ271838
	Chinese wheat mosaic virus RNA2/ Rongcheng (CWMV RNA 2/ Rongcheng)	AJ271839
	Oat golden stripe virus RNA 1 (OGSV RNA 1)	NC_002358
	Oat golden stripe virus RNA 2 (OGSV RNA 2)	NC_002357
	Soil-borne cereal mosaic virus RNA1 (SBCMV RNA 1)	NC_002351
	Soil-borne cereal mosaic virus RNA2 (SBCMV RNA 2)	NC_002330
	Soil-borne wheat mosaic virus RNA 1/ Japanese Northern (SBWMV RNA 1/JN)	AB299271
	Soil-borne wheat mosaic virus RNA2/ Japanese Northern (SBWMV RNA 2/JN)	AB299272
	Soil-borne wheat mosaic virus RNA 1/ US-Nebraska, 1981 wild-type (SBWMV RNA 1. N)	L07937
	Soil-borne wheat mosaic virus RNA 2/ US-Nebraska, 1988 wild-type (SBWMV RNA 2 N)	L07938
	Soil-borne wheat mosaic virus RNA 1 / Wiltshire (SBWMV RNA 1 / Wiltshire)	AJ298068
	Soil-borne wheat mosaic virus RNA 2 / Wiltshire (SBWMV RNA 2 / Wiltshire)	AJ298069
	Sorghum chlorotic spot virus RNA 1 (SrCSV RNA 1)	NC_004014
	Sorghum chlorotic spot virus RNA 2 (SrCSV RNA 2)	NC_004015
Hordeivirus	Barley stripe mosaic virus RNA1/China (BSMV RNA 1/China)	AY789693
	Barley stripe mosaic virus RNA 2/China (BSMV RNA 2/China)	AY789694
	Barley stripe mosaic virus RNA3/China (BSMV RNA 3/China)	AY787207
	Barley stripe mosaic virus RNA1/ ATCC.PV-43 (BSMV RNA 1/ATCC.PV-43)	NC_003469
	Barley stripe mosaic virus RNA 2/ ATCC.PV-43 (BSMV RNA 2/ATCC.PV-43)	NC_003481
	Barley stripe mosaic virus RNA3/ ATCC.PV-43 (BSMV RNA 3/ATCC.PV-43)	NC_003478
	Barley stripe mosaic virus RNA3/unknown isolate (BSMV RNA 3/unknown isolate)	X52774
	Barley stripe mosaic virus RNA 1/Type (BSMV RNA 1/Type)	U35768
	Barley stripe mosaic virus RNA2/Type (BSMV RNA 2/Type)	U35772
	Barley stripe mosaic virus RNA 1.ND18 (BSMV RNA 1/ND18)	U35767
	Barley stripe mosaic virus RNA 2.ND18 (BSMV RNA 2.ND18)	U35770
	Barley stripe mosaic virus RNA 3.ND18 (BSMV RNA 3.ND18 clone pND18 gamma 42)	U13917
	Barley stripe mosaic virus RNA 3.ND18 (BSMV RNA 3.ND18 clone pBSM gamma-a)	M16577
	Barley stripe mosaic virus RNA1.CV42 (BSMV RNA 1.CV42)	U05347
	Barley stripe mosaic virus RNA 2.CV42 (BSMV RNA 2.CV42)	U35771

Genus (in bold) and unclassified viruses	Virus.Isolate	GenBank accession number
	Barley stripe mosaic virus RNA 1.CV17 (BSMV RNA 1.CV17)	U35766
	Barley stripe mosaic virus RNA 2.CV17 (BSMV RNA 2 CV17)	U35769
Pecluvirus	Indian peanut clump virus RNA 2.H (IPCV RNA 2.H)	NC_004730
	Indian peanut clump virus RNA 1. Hyderabad serotype (IPCV RNA 1. Hyderabad serotype)	NC_004729
	Indian peanut clump virus D RNA 2.D (IPCV RNA 2.D)	AF447396
	Indian peanut clump virus L RNA 2 (IPCV RNA 2.L)	AF239729
Pecluvirus	Peanut clump virus RNA 1 (PCV RNA 1)	NC_003672
	Peanut clump virus RNA 2. 87/TGTA2 (PCV RNA 2.87/TGTA2)	NC_003668
	Peanut clump virus Ni RNA 2 (PCV RNA 2.Ni)	AF447398
	Peanut clump virus B RNA 2 (PCV RNA 2.B)	AF447401
	Peanut clump virus M RNA 2 (PCV RNA 2.M)	AF447400
	Peanut clump virus N RNA 2 (PCV RNA 2.N)	AF447399
Pomovirus	Beet soil-borne virus. PL RNA 3 (BSBV RNA 3.PL)	FJ971719
	Beet soil-borne virus. PL RNA 2 (BSBV RNA 2.PL)	FJ971718
	Beet soil-borne virus. PL RNA 1 (BSBV RNA 1.PL)	FJ971717
	Beet soil-borne virus RNA 2.Ahlum (BSBV RNA 2.Ahlum)	U64512
	Beet soil-borne virus. XJ RNA 3 (BSBV RNA 3.XJ)	EF545143
	Beet soil-borne virus. XJ RNA 2 (BSBV RNA 2.XJ)	EF545141
	Beet soil-borne virus. XJ RNA 1 (BSBV RNA 1.XJ)	EF545139
	Beet soil-borne virus. IM RNA 3 (BSBV RNA 3.IM)	EF545142
	Beet soil-borne virus. IM RNA 2 (BSBV RNA 2.IM)	EF545140
	Beet soil-borne virus RNA 1. IM (BSBV RNA 1.IM)	EF545138
	Beet soil-borne virus RNA 1.Ahlum (BSBV RNA 1.Ahlum)	NC_003520
	Beet soil-borne virus RNA 3.Ahlum (BSBV RNA 3.Ahlum)	NC_003519
	Beet soil-borne virus RNA 3. pUBS-3-wt (BSBV RNA 3. pUBS-3-wt)	FN386614
	Beet soil-borne virus RNA 2. pUBS-2-wt (BSBV RNA 2. pUBS-2-wt)	FN386613
	Beet soil-borne virus RNA 1. pUBS-1-wt (BSBV RNA 1. pUBS-1-wt)	FN386612
	Broad bean necrosis virus RNA 3 (BBNV RNA 3)	NC_004425
	Broad bean necrosis virus RNA 2 (BBNV RNA 2)	NC_004424
	Broad bean necrosis virus RNA 1 (BBNV RNA 1)	NC_004423
	Potato mop-top virus RNA 3.Sw (PMTV RNA 3.Sw)	NC_003725
	Potato mop-top virus RNA 1.Sw (PMTV RNA 1.Sw)	NC_003723
	Potato mop-top virus RNA 2.Sw (PMTV RNA 2.Sw)	NC_003724
	Potato mop-top virus RNA 2. Korneta-Nemilkov (PMTV RNA 2. Korneta-Nemilkov)	DQ102381
	Potato mop-top virus RNA3. Korneta-Nemilkov (PMTV RNA 3. Korneta-Nemilkov)	DQ144451
	Potato mop-top virus. RNA 3.Todd (PMTV RNA 3.Todd)	D30753
	Beet virus Q RNA 1 (BVQ RNA 1)	NC_003510
	Beet virus Q RNA 2 (BVQ RNA 2)	NC_003511
	Beet virus Q RNA 3 (BVQ RNA 3)	NC_003512
Tobravirus	Pea early browning virus RNA 1.British SPS (PEBV RNA 1. British SPS)	NC_002036
	Pea early browning virus RNA 2.British SPS (PEBV RNA 2. British SPS)	NC_001368
	Pepper ringspot virus RNA 1.CAM (PepRSV RNA 1.CAM)	NC_003669
	Pepper ringspot virus RNA 2.CAM (PepRSV RNA 2.CAM)	NC_003670
	Tobacco rattle virus RNA1. ppK20 (TRV RNA 1. ppK20)	AF314165
	Tobacco rattle virus RNA 1. ORY (TRV RNA 1.ORY)	AF034622
	Tobacco rattle virus RNA2.ORY (TRV RNA 2.ORY)	AF034621
	Tobacco rattle virus RNA 1. MI-1 (TRV RNA 1.MI-1)	GQ903771
	Tobacco rattle virus RNA2. MI-1 (TRV RNA 2.MI-1)	GQ903772
	Tobacco rattle virus RNA2. OR2 (TRV RNA 2.OR2)	AY166661
Tobravirus	Tobacco rattle virus RNA2. BM (TRV RNA 2.BM)	AY166663
	Tobacco rattle virus RNA2. Cot2 (TRV RNA 2.Cot2)	AY166662
	Tobacco rattle virus RNA2. Umt1 (TRV RNA 2. Umt1)	AY166660

Table 1. (Continued)

Genus (in bold) and unclassified viruses	Virus.Isolate	GenBank accession number
	Tobacco rattle virus RNA2.Potato (TRV RNA 2.Potato)	AB369280
	Tobacco rattle virus RNA 2.K (TRV RNA 2.K)	AJ536414
	Tobacco rattle virus RNA 2.ppK20 (TRV RNA 2.ppK20)	NC_003811
	Tobacco rattle virus RNA 1.SYM (TRV RNA 1.SYM)	X06172
	Tobacco rattle virus RNA 2 TCM (TRV RNA 2.TCM)	X03955
	Tobacco rattle virus RNA 1.PpO85 (TRV RNA 1. PpO85)	AJ586803
	Tobacco rattle virus RNA 2.Rostock (TRV RNA 2.Rostock)	AJ272198
	Tobacco rattle virus RNA 2.SP (TRV RNA 2.SP)	AJ007293
Tobamovirus	Brugmansia mild mottle virus.2373 (BruMMV.2373)	NC_010944
	Cucumber fruit mottle mosaic virus (CFMMV)	NC_002633
	Cucumber green mottle mosaic virus. CCMV_TANG (CGMMV. CCMV_TANG)	HM008919
	Cucumber green mottle mosaic virus. BG (CGMMV.BG)	FJ654659
	Cucumber green mottle mosaic virus. C (CGMMV)	FJ654658
	Cucumber green mottle mosaic virus. VIROG-43M (CGMMV. VIROG-43M)	GQ495275
	Cucumber green mottle mosaic virus. MC-2 (CGMMV. MC-2)	GQ495274
	Cucumber green mottle mosaic virus. SP. Alm08 (CGMMV. SP.Alm08)	GQ411361
	Cucumber green mottle mosaic virus GX (CGMMV.GX)	GQ277655
	Cucumber green mottle mosaic virus.India-Delhi (CGMMV. India-Delhi)	DQ767631
	Cucumber green mottle mosaic virus. LN (CGMMV.LN)	EU352259
	Cucumber green mottle mosaic virus. KOM (CGMMV.KOM)	AF417243
	Cucumber green mottle mosaic virus. KW (CGMMV.KW)	AF417242
	Cucumber green mottle mosaic virus. BG-SB ((CGMMV. BG-SB)	FJ654657
	Cucumber green mottle mosaic virus. MC-1 (CGMMV.MC-1)	FJ848666
	Cucumber green mottle mosaic virus.SH (CGMMV.SH)	NC_001801
	Cucumber green mottle mosaic virus.Watermelon (CGMMV.Watermelon)	AB369274
	Cucumber green mottle mosaic virus. Liaoning (CGMMV. Liaoning)	EF611826
	Hibiscus latent Singapore virus (HLSV)	NC_008310
	Kyuri green mottle mosaic virus.KGMMV-C1 (KGMMV-C1)	NC_003610
	Kyuri green mottle mosaic virus.Yodo (KGMMV.Yodo)	AB015145
	Kyuri green mottle mosaic virus.YM (KGMMV.YM)	AB162006
	Obuda pepper virus.Ob (ObPV.Ob)	NC_003852
	Odontoglossum ringspot virus. Taiwan (ORSV.Taiwan)	AY571290
	Odontoglossum ringspot virus (unknown isolate)(ORSV)	U89894
	Odontoglossum ringspot virus.Cy-1 (ORSV.Cy-1)	S83257
	Odontoglossum ringspot virus. Singapore-1 (ORSV. Singapore-1)	U34586
	Odontoglossum ringspot virus.18kDa (ORSV. 18kDa)	NC_001728
	Paprika mild mottle virus.Japanese (PaMMV.Japanese)	NC_004106
Tobamovirus	Pepper mild mottle virus.L4BV (PMMoV.L4BV)	AB276030
	Pepper mild mottle virus.S (PMMoV.S)	NC_003630
	Pepper mild mottle virus.Iw. (PMMoV.Iw)	AB254821
	Pepper mild mottle virus.CN (PMMoV.CN)	AY859497
	Rehmannia mosaic virus.Henan (RehMV.Henan)	NC_009041
	Ribgrass mosaic virus. Impatiens (RMV.Impatiens)	DQ223770
	Ribgrass mosaic virus. Shanghai (RMV.Shanghai)	AF254924
	Streptocarpus flower break virus (SFBV)	NC_008365
	Tobacco mild green mosaic virus. U2 (small type). DSMZ. PV-0113 (TMGMV. U2. PV-0113	EF469769
	Tobacco mild green mosaic virus. HP (TMGMV.HP)	DQ821941
	Tobacco mild green mosaic virus (Unknown isolate)(TMGMV)	NC_001556
	Tobacco mild green mosaic virus.Japanese (TMGMV.Japanese)	AB078435

Genus (in bold) and unclassified viruses	Virus.Isolate	GenBank accession number
	Tobacco mosaic virus. TMV-017 (TMV-017)	AF395128
	Tobacco mosaic virus (unknown isolate 1) (TMV-1)	AF155507
	Tobacco mosaic virus isolate.152 (TMV.152)	AF395129
	Tobacco mosaic virus pet-TW (TMV. pet-TW)	EF392659
	Tobacco mosaic virus. Fujian (TMV.Fujian)	AF395127
	Tobacco mosaic virus (unknown isolate 2) (TMV-2)	NC_001367
	Tobacco mosaic virus. IM (TMV.IM)	AB369276
	Tobacco mosaic virus. Pet (TMV.Pet)	AB369275
	Tobacco mosaic virus.Crucifer (TMV.Crucifer)	Z29370
	Tobacco mosaic virus. K2- Kazakhstan strain (TMV. K2-Kazakhstan strain)	Z92909
	Tobacco mosaic virus. ssp. NC 82. Korean common strain (TMV. ssp. NC 82)	X68110
	Tobacco mosaic virus K1. Kazakh strain (TMV. K1. Kazakh strain)	AJ243571
	Tobacco mosaic virus. B935A (TMV. B935A)	AJ011933
	Tomato mosaic virus. XJT-1 (ToMV. XJT-1)	FN985165
	Tomato mosaic virus. Queensland (ToMV. Queensland)	NC_002692
	Tomato mosaic virus.NS (ToMV.NS)	GQ280794
	Tomato mosaic virus.L11Y (ToMV.L11Y)	AB355139
	Tomato mosaic virus. ToMV1-2 (ToMV1-2)	DQ873692
	Tomato mosaic virus. Camellia (ToMV.Camellia)	AJ417701
	Tomato mosaic virus.L11A-Fukushima (ToMV. L11A-Fukushima)	AB083196
	Turnip vein-clearing virus.OSU (TVCV. OSU)	NC_001873
	Youcai mosaic virus (YMoV)	U30944
	Youcai mosaic virus. 722 (YMoV.722)	AB261175
	Youcai mosaic virus. crucifer strain Cg (YMoV. crucifer strain Cg)	D38444
	Chinese Rape Mosaic Virus. Wh (CRMV.Wh)	EU571218
Unclassified *Tobamovirus*	Zucchini green mottle mosaic virus. K (ZGMMV.K)	NC_003878
	Zucchini green mottle mosaic virus. ZT-1 (ZGMMV. ZT-1)	AJ252189
	Abutilon yellow mosaic virus (AYMV)	EU559678
	Bell pepper mottle tobamovirus (BPeMV)	DQ355023
	Cactus mild mottle virus.Kr (CMMoV.Kr)	EU043335
	Crucifer tobamovirus (CV)	AY318866
Unclassified *Tobamovirus*	Crucifer tobamovirus. Wasabi. Shizuoka (CV. Wasabi. Shizuoka)	NC_003355
	Crucifer tobamovirus. Wasabi. Tochigi (CV. Wasabi. Tochigi)	AB017504
	Cucumber mottle virus (CMoV)	NC_008614
	Maracuja mosaic virus (MarMV)	DQ356949
	Penstemon ringspot virus (PRSV)	NC_008295

In the parsimony method, the total numbers of synonymous (c_S) and nonsynonymous (c_N) substitutions as well as the average numbers of synonymous (s_S) and nonsynonymous (s_N) sites per codon over the phylogenetic tree for each codon site were computed according to the maximum parsimony principle (Fitch 1971; Hartigan 1973). The null hypothesis of selective neutrality (r_S=r_N or ω = 1) was tested for each site by computing the probability (p) of obtaining the observed or more biased values for c_S and c_N, which were assumed to follow a binomial distribution with the probabilities of occurrence of synonymous and nonsynonymous substitutions given by $s_S/(s_S + s_N)$ and $s_N/(s_S + s_N)$, respectively. Positive selection is inferred when $p<0.05$ and $c_N/s_N>c_S/s_S$ (Suzuki 2006).

Potential recombination events between diverged nucleotide sequences were explored with two programs: RDP version 3.31β (Martin et al. 2005b) and RECCO (Maydt and

Lengauer 2006). RDP incorporates several published recombination detection methods into a single suite of tools: RDP (Martin and Rybicki 2000), GENECONV (Padidam et al. 1999), BOOTSCAN (Martin et al. 2005a), MAXCHI (Smith 1992), CHIMAERA (Posada and Crandall 2001), SISCAN (Gibbs et al. 2000), and 3SEQ (Boni et al. 2007). In all cases, defaults parameters were used. Only events predicted by more than half of the methods are considered as significant. The algorithm developed and described by Maydt and Lengauer (2006) as fast, simple and sensitive method for detecting recombination in a set of sequences and locating putative recombination breakpoints is based on cost minimization. This method has only two tunable parameters, recombination and mutation cost. In practice the only parameter considered is α, representing the cost of mutation relative to recombination. When α changes from 0 to 1, the cost of mutation weighted by α increases, and the cost for recombination weighted by 1- α decreases. In other words the parameter α controls the ambiguity between mutation and recombination.

Results

Recombination Events during Evolution of the *Virgaviridae*

Examination of the RECCO program output regarding the occurrence of recombination events in the entire genome of the *Virgaviridae* revealed that all virus genera composing this family were constituted by potential recombinants. In fact, the analyses showed that 75% (12/16), 59% (10/17), 90% (9/10), 81% (22/27), 81% (17/21), and 44% (34/77) of members of *Furovirus*, *Hordeivirus*, *Pecluvirus*, *Pomovirus*, *Tobravirus* and *Tobamovirus* genera were putative recombinants, respectively. RDP 3.31β algorithm confirmed these results only for *Pecluvirus*, *Pomovirus*, *Tobravirus*, and *Tobamovirus* genera with the following percentages (ratios): 30% (3/10), 7.5% (2/27), 38% (8/21), and 0.5% (4/77), respectively (Table 2). In *Furovirus* genus, recombination sites occurred in RNA 1 as well as in RNA 2 segments. Obviously the RNA 1 genomic portion, being longer than RNA 2, carried the greater number of recombination breakpoints. It has shown that OGSV was the most frequently recombining virus followed by SrCSV, and SBWMV; on the contrary, CWMV, SBCMV, and SBWMV were the lesser frequently recombining viruses. Moreover, in 83% (10/12) of putative recombinant accessions, swapped fragment had a size exceeding three nucleotides. Conversely, in *Hordeivirus* genus, the presence of possible recombination sites was scarce in the three genomic fragments (RNAs 1, 2, and 3). Nevertheless, while in *Furovirus* genus, the breakpoint length did not overstep 53 nucleotides (SrCSV RNA 2), it was as long as 507 nucleotides (BSMV RAN 2 CV 17) in *Hordeivirus* genus. On the other hand, compared to *Furovirus* genus, members of *Pecluvirus* genus had a similar behavior in terms of detected recombination sites, breakpoint length and recombination frequency, but clearly distinct from viruses of *Hordeivirus* genus. According to RDP package, genetic material exchange arose between the two virus species (IPCV and PCV) forming the genus *Pecluvirus*. In *Pomovirus* genus, possible recombination signals occurred in all RNA segments. In about 55% (15/27) of investigated accessions, recombination events were scarce.

Table 2. Determination of inferred putative recombination events and their frequencies along the sequences of the whole genome of members of the family *Virgaviridae*

Genus	Recombination determined by RECCO								Recombination determined by RDP v3.31β
	Virus.isolate			Length of breakpoint				Genomic position of the longest breakpoint (size in nucleotide)	putative parental (Major x Minor)
		NRS	GIRE (nt)	1 residue	2 residues	3 residues	>3residues		
Furovirus	CWMV RNA 1.Rongcheng	1	6814-6836	0	0	0	1	6814-6836 (23)	/
	CWMV RNA 2. Rongcheng	2	6218-6282	0	0	0	2	6218-6264 (47)	/
	OGSV RNA 1	74	312-7165	19	8	15	32	1929-1946 (18)	/
	OGSV RNA 2	20	3409-7154	8	1	2	9	6083-6095 (13)	/
	SBCMV RNA 1	2	31-7286	0	0	1	1	7241-7286 (46)	/
	SBCMV RNA 2	2	3382-7286	0	0	0	2	7241-7286 (46)	/
	SBWMV RNA 1. N	26	228-6994	6	5	6	9	5511-5525 (15)	/
	SBWMV RNA 2. N	16	3386-6857	4	6	3	3	5351-5363 (13)	/
	SBWMV RNA 1. Wiltshire	2	31-7286	0	1	0	1	7241-7286 (46)	/
	SBWMV RNA 2. Wiltshire	2	3385-7286	1	0	0	1	7241-7286 (46)	/
	SrCSV RNA 1	57	58-6962	9	12	11	25	804-818 (15)	/
	SrCSV RNA 2	36	3414-6902	12	5	4	15	3414-3466 (53)	/
Hordeivirus	BSMV RNA 1.China	7	1747-3631	0	0	1	6	3588-3631 (44)	/
	BSMV RNA 2.China	3	2977-3610	0	0	1	2	2977-3007 (31)	/
	BSMV RNA 3.China	4	937-3609	0	0	1	3	937-1189 (253)	/
	BSMV RNA 1/ATCC.PV-43	1	3610	1	0	0	0	3610 (1)	/
	BSMV RNA 3/ATCC.PV-43	1	365-366	0	1	0	0	365-366 (2)	/
	BSMV RNA 3.ND18 clone pBSM gamma-a	1	3588-3604	0	0	0	1	3588-3604 (17)	/
	BSMV RNA 1.CV42	1	2455-2456	0	1	0	0	2455-2456 (2)	/
	BSMV RNA 2.CV42	1	3588-3607	0	0	0	1	3588-3607 (20)	/
	BSMV RNA 1.CV17	1	3590-3610	0	0	0	1	3590-3610 (21)	/
Hordeivirus	BSMV RNA 2 CV17	12	483-3610	1	1	0	10	2032-2538 (507)	/
Pecluvirus	IPCV RNA 2.H	28	1274-3704	11	5	3	9	1971-1993 (23)	/
	IPCV RNA 1. Hyderabad serotype	3	2023-5751	2	0	0	1	5747-5751 (5)	/

Table 2. (Continued)

Genus	Recombination determined by RECCO								Recombination determined by RDP v3.31β
	Virus.isolate			Length of breakpoint				Genomic position of the longest breakpoint (size in nucleotide)	putative parental (Major x Minor)
		NRS	GIRE (nt)	1 residue	2 residues	3 residues	>3residues		
	IPCV RNA 2.D	21	1274-5777	4	3	1	13	1408-1421 (14)	IPCV RNA 2.H x PCV RNA 2.87/TGTA2
	IPCV RNA 2.L	35	901-5816	11	5	5	14	1238-1259 (22)	/
	PCV RNA 1	4	2023-5761	2	1	1	0	5759-5761 (3)	/
	PCV RNA 2.87/TGTA2	27	1018-3901	9	4	5	9	1018-1064 (47)	IPCV RNA 2.L x PCV RNA 2.B
	PCV RNA 2.Ni	14	1256-3844	2	4	2	6	3722-3750 (29)	IPCV RNA 2.L x PCV RNA 2.N)
	PCV RNA 2.B	27	953-5480	5	4	4	14	4927-4946 (20)	/
	PCV RNA 2.M	39	1181-5389	14	12	5	8	5048-5067 (20)	/
	PCV RNA 2.N	21	1472-5638	6	4	6	5	5365-5383 (19)	/
Pomovirus	BSBV RNA 3.PL	2	3095-3179	0	0	0	2	3161-3179 (19)	/
	BSBV RNA 2.PL	2	4273-5421	0	0	0	2	5379-5421 (43)	/
	BSBV RNA 1.PL	1	1643-1673	0	0	0	1	1643-1673 (31)	/
	BSBV RNA 1.XJ	1	4163-4221	0	0	0	1	4163-4221 (59)	/
	BSBV RNA 3.IM	3	2955-4990	0	0	1	2	4867-4960 (94)	/
	BSBV RNA 2.IM	2	4426-5635	0	0	0	2	4426-4566 (141)	/
	BSBV RNA 1.IM	4	1583-4221	0	0	0	4	2218-2313 (96)	/
	BSBV RNA 1.Ahlum	1	5908-5909	1	0	0	0	5908-5909 (1)	/
	BSBV RNA 3.Ahlum)	1	5189-5206	0	0	0	1	5189-5206 (18)	/
	BSBV RNA 3. pUBS-3-wt	1	3295-3312	0	0	0	1	3295-3312 (18)	/
	BSBV RNA 1. pUBS-1-wt	1	1583-1673	0	0	0	1	1583-1673 (91)	/
Pomovirus	BBNV RNA 3	31	2769-5653	8	5	4	14	2844-2965 (122)	/
	BBNV RNA 2	75	2670-5646	33	15	13	14	4554-4569 (16)	/
	BBNV RNA 1	64	148-5883	19	9	15	21	2228-2240 (13)	/
	PMTV RNA 3.Sw	1	5426-5483	0	0	0	1	5426-5483 (58)	BSBV RNA 3.IM x PMTV RNA 2. Korneta-Nemilkov
	PMTV RNA 1.Sw	44	108-6153	16	13	4	11	431-442 (12)	/
	PMTV RNA 2.Sw	1	2486-2489	0	0	0	1	2486-2489 (4)	/

Genus	Recombination determined by RECCO								Recombination determined by RDP v3.31β
	Virus.isolate			Length of breakpoint				Genomic position of the longest breakpoint (size in nucleotide)	putative parental (Major x Minor)
		NRS	GIRE (nt)	1 residue	2 residues	3 residues	>3residues		
	PMTV RNA 2. Korneta-Nemilkov	1	2488	1	0	0	0	2488 (1)	/
	PMTV RNA 3. Korneta-Nemilkov	1	2769	1	0	0	0	2769 (1)	BSBV RNA 3.IM x PMTV RNA 2. Korneta-Nemilkov
	BVQ RNA 1	40	128-5808	11	11	6	12	4858-4875 (18)	/
	BVQ RNA 2	26	2478-5671	7	7	2	10	4761-4771 (11)	/
	BVQ RNA 3	29	2729-5610	13	3	5	8	3930-3945 (17)	/
Tobravirus	PEBV RNA 1. British SPS	35	115-7146	7	7	9	12	6835-6875 (41)	PepRSV RNA 1.CAM x TRV RNA 1.ORY
	PEBV RNA 2. British SPS	66	334-3916	34	13	7	12	2238-2247 (10)	TRV RNA 2.Potato x TRV RNA 2.TCM
	PepRSV RNA 1.CAM	39	139-7108	19	8	1	11	212-218 (7)	/
	PepRSV RNA 2.CAM	30	419-2751	14	3	5	8	2285-2302 (18)	TRV RNA 2.SP x TRV RNA 2.ORY
	TRV RNA 2.ORY	1	3730-3744	0	0	0	1	3730-3744 (15)	/
	TRV RNA 1.MI-1	5	182-4330	1	0	0	4	1507-1638 (132)	/
Tobravirus	TRV RNA 2.MI-1	8	968-4391	0	1	1	6	968-1116 (149)	TRV RNA 2.K x TRV RNA 2.Potato
	TRV RNA 2.OR2	3	2640-3996	0	0	1	2	2698-2764 (67)	TRV RNA 2.ORY x TRV RNA 2.BM
	TRV RNA 2.BM	1	2447-2450	0	0	0	1	2447-2450 (4)	TRV RNA 2.K x TRV RNA 2.Potato
	TRV RNA 2. Umt1	3	3068-4061	2	0	0	1	3995-4061 (7)	/
	TRV RNA 2.Potato	3	3145-3996	1	1	0	1	3145-3151 (7)	/
	TRV RNA 2.K	1	3744	1	0	0	0	3744 (1)	/
	TRV RNA 2.ppK20	74	676-4099	27	20	7	20	879-909 (31)	/
	TRV RNA 2.TCM	36	542-3814	17	11	2	6	665-674 (10)	TRV RNA 2.SP x TRV RNA 2.MI-1
	TRV RNA 1. PpO85	3	182-4502	1	1	0	1	1507-1528 (22)	/
	TRV RNA 2.Rostock	20	454-2772	7	7	3	3	454-495 (42)	/
	TRV RNA 2.SP	40	359-4190	16	9	2	13	2197-2215 (19)	TRV RNA 2.ppK20 x TRV RNA 2.MI-1

Table 2. (Continued)

Genus	Recombination determined by RECCO								Recombination determined by RDP v3.31β
	Virus.isolate			Length of breakpoint				Genomic position of the longest breakpoint (size in nucleotide)	putative parental (Major x Minor)
		NRS	GIRE (nt)	1 residue	2 residues	3 residues	>3residues		
Tobamovirus	BruMMV.2373	95	58-6965	30	17	14	34	3445-3491 (47)	/
	CFMMV	58	22-6997	16	7	7	28	1018-1061 (44)	/
	CGMMV.BG	1	6139-6257	0	0	0	1	6139-6257 (119)	/
	CGMMV	2	2501-7002	1	0	0	1	2501-2520 (20)	/
	CGMMV.GX	3	76-6974	1	0	0	2	1561-1875 (315)	/
	CGMMV. India-Delhi	2	2521-2932	0	0	0	2	2700-2932 (233)	/
	CGMMV.LN	2	3249-6938	1	0	0	1	3249-3527 (279)	/
	CGMMV.KOM	1	4887-4889	0	0	1	0	4887-4889 (3)	/
	CGMMV. BG-SB	2	1467-1641	0	0	0	2	1467-1547 (81) 1561-1641 (81)	/
	CGMMV.SH	2	2700-3181	0	0	0	2	2700-2932 (233)	/
Tobamovirus	HLSV	39	146-6973	14	4	9	12	6470-6476 (7)	/
	KGMMV-C1	15	1600-4913	2	2	1	10	2348-2368 (21)	/
	KGMMV.Yodo	8	1876-3210	2	0	3	3	2333-2360 (28)	KGMMV.YM x CFMMV
	ObPV.Ob	6	4657-6742	2	2	1	1	5268-5271 (4)	/
	ORSV. Singapore-1	1	5144-5503	0	0	0	1	5144-5503 (360)	YMoV x BPeMV
	ORSV. 18kDa	1	3460-3990	0	0	0	1	3460-3990 (531)	YMoV x BPeMV
	PaMMV.Japanese	8	4668-6742	4	1	1	2	4668-4676 (9)	/
	RehMV.Henan	28	299-6629	2	3	4	19	3840-3897 (58)	/
	RMV.Impatiens	2	4734-5532	0	0	0	2	5297-5532 (236)	/
	SFBV	120	18-6935	44	24	26	26	6725-6772 (48)	/
	TMV.152	1	5434-6385	0	0	0	1	5434-6385 (952)	/
	TMV. pet-TW	15	24-6999	5	0	2	8	915-1205 (291)	/
	ToMV. XJT-1	1	3240-3991	0	0	0	1	3240-3991 (752)	/
	ToMV. Queensland	1	3381-3434	0	0	0	1	3381-3434 (54)	/
	ToMV.NS	3	2718-6075	0	0	0	3	3240-3420 (181)	/
	ToMV.L11Y	1	3435-3664	0	0	0	1	3435-3664 (230)	/

Genus	Recombination determined by RECCO								Recombination determined by RDP v3.31β
	Virus.isolate			Length of breakpoint				Genomic position of the longest breakpoint (size in nucleotide)	putative parental (Major x Minor)
		NRS	GIRE (nt)	1 residue	2 residues	3 residues	>3residues		
Tobamovirus	TVCV. OSU	3	130-6983	0	1	0	2	6868-6983 (116)	/
	CRMV.Wh	7	2012-5710	2	0	0	5	5385-5532 (148)	/
	AYMV	27	555-6837	10	5	6	6	6470-6837 (368)	/
	BPeMV	40	313-6660	7	3	9	21	4848-4871 (24)	/
	CMMoV.Kr	180	22-6963	49	33	30	68	1458-1485 (28)	/
	CMoV	100	20-6709	26	23	19	32	591-615 (25)	/
	MarMV	153	52-7012	40	26	31	56	5956-5978 (23)	KGMMV.YM x ToMV. XJT-1
	PRSV	2	128-6995	1	0	1	0	128-130 (3)	/

With RDP v3.31β algorithm, only events supported by more than half of the different methods are reported. Nucleotide numbering corresponds to the aligned sequences. Abbreviations: NRS: number of recombination sites, GIRE: genomic interval of recombination events (the span of sequences in the genome of the virus where recombination events were predicted).

In the remaining accessions, recombination breakpoints were more frequent. Indeed, investigations conducted in the genome of BBNV and BVQ showed that the number of putative recombination sites was between 26 (BVQ RNA 2) and 75 (BBNV RNA 2). In 74% (20/27) of screened accessions, breakpoint lengths exceeded three nucleotides; the longest was 141 residues for BSBV RNA. IM. In *Tobravirus* genus, all virus components were potential recombinants. Although PEBV and PepRSV showed substantial recombination signals (between 30 and 66), 61% (8/13) of screened TRV accessions recombined in positions whose number was between one and eight, and 39% (5/13) of them had a number of putative recombination sites close to those of the former two tobraviruses. On recombination frequency, the majority of recombination intervals consisted of one, two and more than three nucleotides. While the longest breakpoint length was 149 nucleotides (TRV RNA 2 MI.1), the shortest was a single residue (TRV RNA 2.K). In *Tobamovirus* genus, 75% (21/28) of typical tobamoviruses showed a poor frequency of potential recombination events (no more than eight putative sites), whereas in the remaining analyzed accessions, SFBV was the most frequently recombining virus with as many as 120 possible breakpoints followed by BruMMV.2373, CFMMV, HLSV, and RehMV. Henan. On recombination frequency, most of recombination intervals exceeded three nucleotides. Those were in some instances as long as 952 nucleotides (TMV.152). By contrast to all examined accessions in the *Virgaviridae*, the greatest number of putative breakpoints was detected in the unclassified tobamovirus CMMoV.Kr (180 positions) followed by MarMV (153 positions), and CMoV (100 positions). The recombination intervals consisted of one, two, three as well as more than three nucleotides. In unclassified tobamovirus group, the length of exchanged segment was between three and 368 nucleotides (Table 2).

Nucleotide Sequence Analysis

Maximum composite likelihood estimate of the pattern of nucleotide substitution were conducted in MEGA4.1β. The studied parameters were the following: the rates of different transitional and transversional substitutions, the nucleotide frequencies, the transition/transversion rate ratios including 2 coefficients k_1 (purines) and k_2 (pyrimidines), the overall transition /transversion bias R where $R = [AGk_1 + TCk_2]/[(A+G)(T+C)]$, and the number of total of positions in the final dataset. . In all these analyses, codon positions included were first + second + third + noncoding. All positions containing gaps and missing data were eliminated from the dataset (complete deletion option). The results for each virus genus are shown in Table 3.

The MEGA4.1β program implements also the Tajima's Neutrality Test. The purpose of this test is to identify sequences which do not fit the neutral theory model at equilibrium between mutation and genetic drift. Tajima's test compares a standardized measure of the total number of segregating sites (polymorphic DNA sites) in the sampled DNA and the average number of mutations between pairs in the sample. The Tajima's D was determined for all analyzed sequences (Table 3).

Table 3. Description of Maximum Composite Likelihood estimate of the pattern of nucleotide substitution in the entire genome of each genus in the *Virgaviridae* and related parameters used

Nucleotide substitution pattern					Tajima's *D*	NF	*k*1	*k*2	R
	A	T	C	G			2.218	2.993	1.117
A	-	*6.26*	*3.86*	12.59	3.328273	0.277			
T	*6.05*	-	11.56	*5.68*		0.287			
C	*6.05*	18.74	-	*5.68*		0.177			
G	13.41	*6.26*	*3.86*	-		0.260			
	A	T	C	G			16.36	16.865	6.960
A	-	*1.53*	*1.03*	20.31	3.237907	0.292			
T	*1.57*	-	17.36	*1.24*		0.285			
C	*1.57*	25.88	-	*1.24*		0.191			
G	25.70	*1.53*	*1.03*	-		0.231			
	A	T	C	G			2.248	2.605	1.094
A	-	*6.60*	*4.18*	13.18	2.390468	0.264			
T	*5.98*	-	10.90	*5.86*		0.292			
C	*5.98*	17.20	-	*5.86*		0.185			
G	13.45	*6.60*	*4.18*	-		0.259			
	A	T	C	G			27.485	31.198	12.164
A	-	*0.94*	*0.54*	21.99	4.782511	0.289			
T	*0.92*	-	16.79	*0.80*		0.294			
C	*0.92*	29.39	-	*0.80*		0.168			
G	25.42	*0.94*	*0.54*	-		0.250			
	A	T	C	G			19.215	22.833	8.875
A	-	*1.29*	*0.76*	20.71	3.354188	0.282			
T	*1.23*	-	17.44	*1.08*		0.296			
C	*1.23*	29.53	-	*1.08*		0.175			
G	23.60	*1.29*	*0.76*	-		0.247			
	A	T	C	G			1.527	57.795	11.098
A	-	*0.98*	*0.60*	1.20	4.540576	0.279			
T	*0.91*	-	34.43	*0.79*		0.298			
C	*0.91*	56.43	-	*0.79*		0.182			
G	1.40	*0.98*	*0.60*	-		0.240			

Rates of different transitional substitutions are shown in bold and those of transversionsal substitutions are shown in italics. Tajima's *D* values are also indicated. Abbreviations: NF: Nucleotide frequency, k1: coefficient (purines), k2: coefficient (pyrimidines), R: transition/transversion bias, NP: number of total positions.

Table 4. Accession pairs of the *Virgaviridae* (*Pomovirus* and *Tobamovirus* genera) under positive selection with determined probability at the 5% level based on the Fisher's Exact Test of Neutrality and calculated d_N-d_S

Genus	Accession pairs	Virus.isolate pairs	*P* value at 5% level	d_N-d_S
Pomovirus	EF545139/DQ102381	BSBV RNA 1.XJ/ PMTV RNA 2. Korneta-Nemilkov	0.034	1.728
	EF545139/NC_003724	BSBV RNA 1.XJ/ PMTV RNA 2.Sw	0.035	1.735
	FJ971717/DQ102381	BSBV RNA 1.PL/ PMTV RNA 2. Korneta-Nemilkov	0.026	1.804
	FJ971717/ NC_003724	BSBV RNA 1.PL/ PMTV RNA 2.Sw	0.027	1.811
	NC_003520/ DQ102381	BSBV RNA 1.Ahlum// PMTV RNA 2. Korneta-Nemilkov	0.035	1.760
	NC_003520/ NC_003724	BSBV RNA 1.Ahlum/ PMTV RNA 2.Sw	0.037	1.766
Tobamovirus	DQ356949/AB355139	MarMV/ ToMV.L11Y	0.041	1.688

Positive Selection

The high genetic stability of viruses can be attributed to negative or purifying selection to maintain the functional integrity of the viral genome. The degree of negative selection in genes, or the degree of functional constraint for the maintenance of the encoded protein sequence, can be estimated, as indicated above, by the ratio between the nucleotide diversities in nonsynonymous and synonymous positions (d_N/d_S). For most coding genes the d_N/d_S ratio is below unity which is consistent with negative selection against protein change.

In contrast, a d_N/d_S ratio above unity may be an indication that adaptive or positive selection is driving gene divergence.

In this study, pairwise comparisons of all screened accessions within each genus showed that, only a few members of the *Pomovirus* genus and one member of *Tobamovirus* genus along with an unclassified tobamovirus were under positive selection.

In fact, in the first, six accession pairs involving five lineages were identified and represented by the following viruses: BSBV, and PMTV; whereas in the second and the third, two lineages representing the virus ToMV and the virus MarMV, were also detected (Table 4).

All these results were obtained by testing the neutrality in sequence pairs with Fisher's Exact Test. The probability of rejecting the null hypothesis of strict-neutrality ($d_N = d_S$) in favor of positive selection for each sequence pair was determined. Values of *P* less than 0.05 were considered significant at the 5% level. The variance of the difference (d_N-d_S) was computed using the bootstrap method (500 replicates).

All analyses were conducted using the Nei-Gojobori method incorporated in MEGA program. All positions containing gaps and missing data were eliminated from the dataset (complete deletion option). There were a total of 329 and 1019 codons in the final dataset for *Pomovirus* and *Tobamovirus* genera, respectively.

Phylogenetic Relationships

Due to its high heterogeneity, the family *Virgaviridae* encompasses divergent virus genera. Thus, inter-genera analyses led to inconclusive results. Consequently, inferred phylogenies were processed within each genus taken separately. The phylogenetic relationships among members of each genus, based on the sequences of their complete genome, were inferred using NJ, ME, MP, ML, and UPGMA.

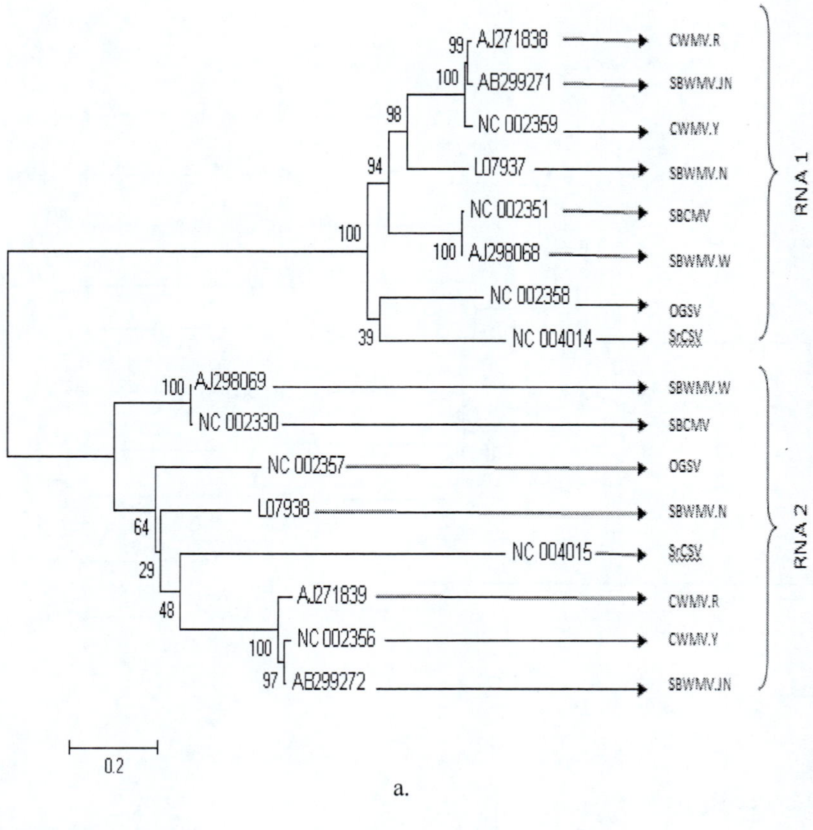

a.

Figure 1. (Continued).

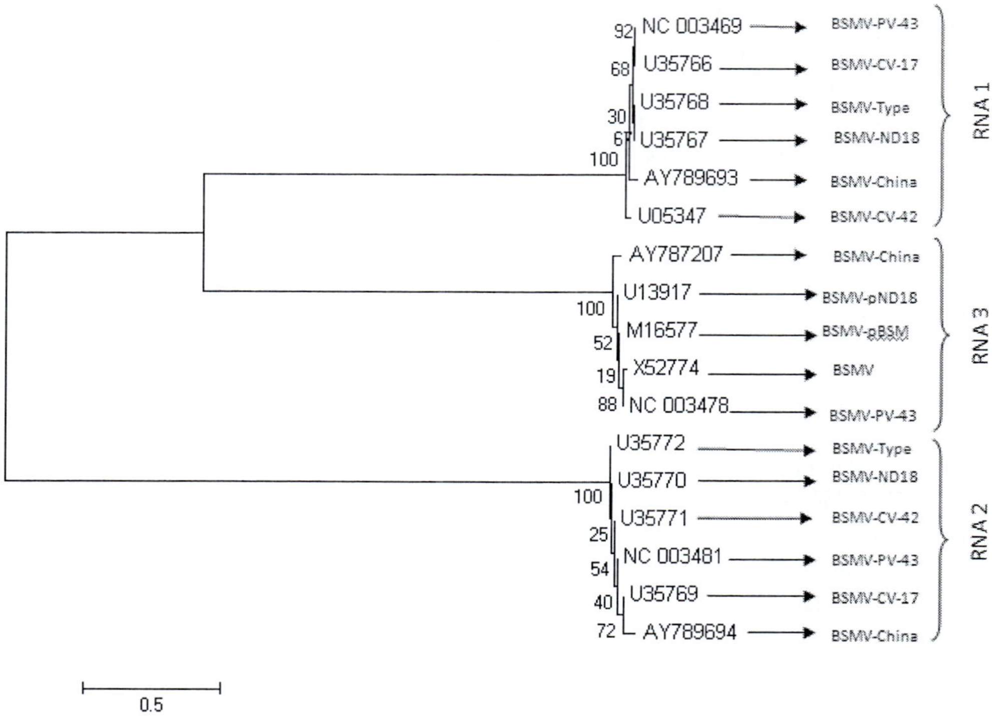

b.

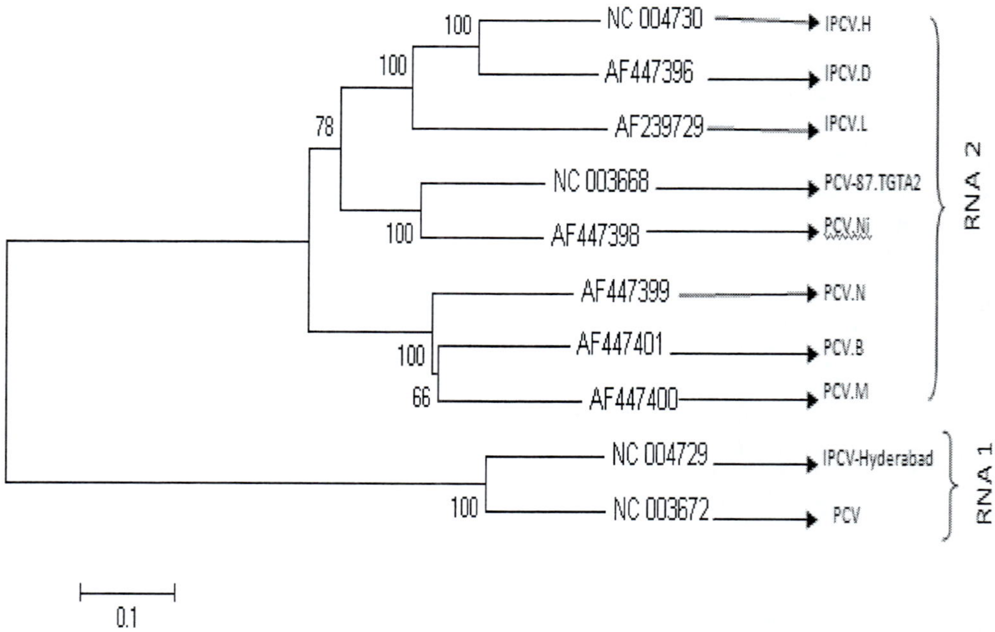

c.

Figure 1. (Continued).

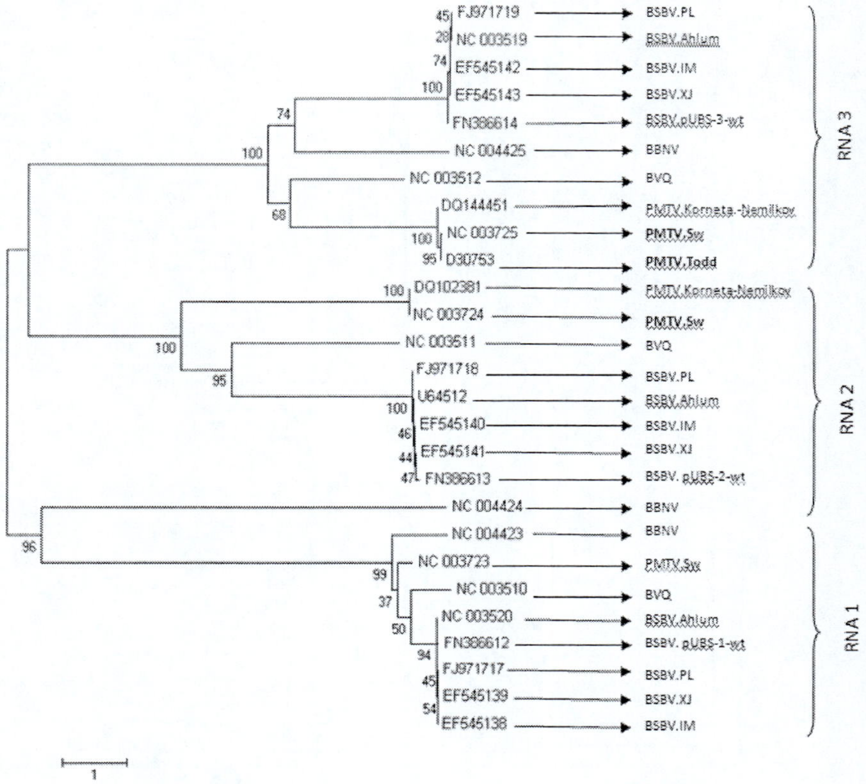

d.

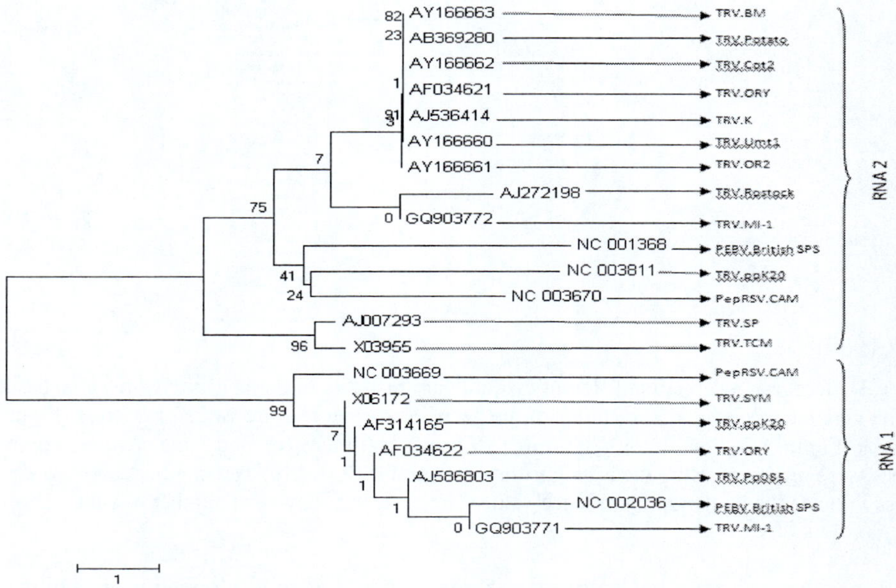

e.

Figure 1. (Continued).

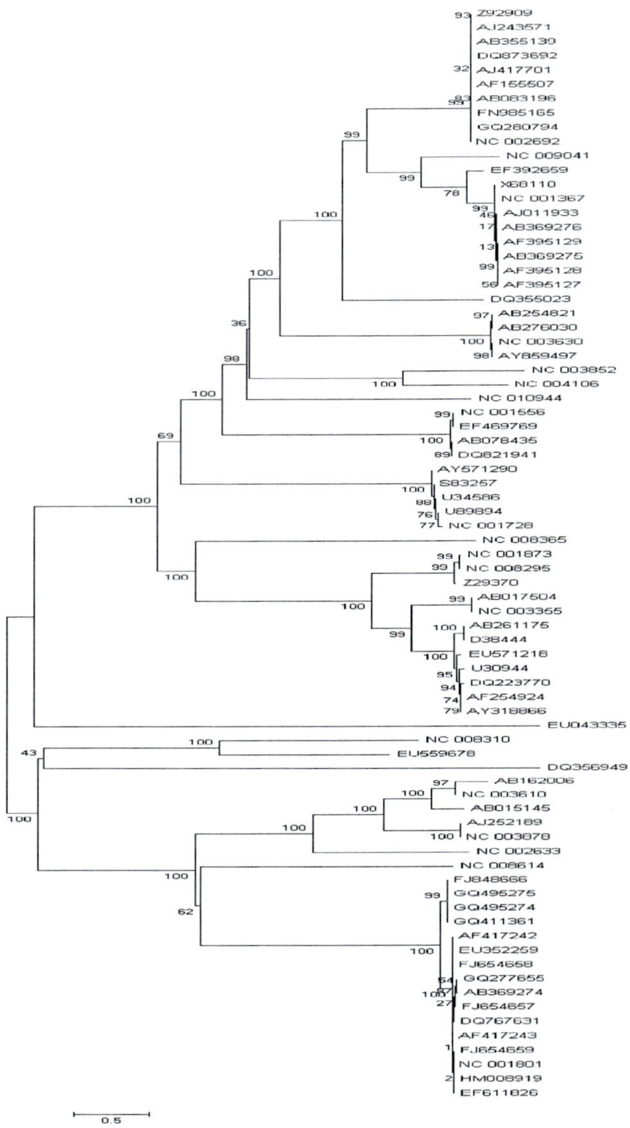

f.

Figure 1. Dendrograms depicting different evolutionary histories of the family *Virgaviridae* based on the entire genome sequences of members of the genus *Furovirus* (Figure 1a), *Hordeivirus* (Figure 1b), *Pecluvirus* (Figure 1c), *Pomovirus* (Figure 1d), *Tobravirus* (Figure 1e), and *Tobamovirus* (Figure 1f). The trees were produced using the NJ algorithm option of MEGA4.1β. The numbers above the branches indicate the bootstrap confidence value. The scale bar shows the number of substitutions per nucleotide.

The constructed trees basically gave a somewhat similar phylogenetic grouping for the five methods used in this study. As indicated in Figure1a,b,c,d,e, the topology of reconstructed trees showed that virus isolates within the genera *Furovirus*, *Hordeivirus*, *Pecluvirus*, *Pomovirus*, and *Tobravirus* segregated into two, three, two, three, and two clusters, respectively. These clades were completely concordant with their genomic segments. According to Figure1f, the phylogenetic reconstruction demonstrated that TMV isolates diverged and split into three distant groupings. Although the first group encompassed the isolates: TMV.Kazakhstan, and TMV.China, the second and the third were formed by isolates described from Korea, China, and Taiwan, and an isolate found on crucifers from Russia. It is worth noting that in the first group, TMV and ToMV isolates were phylogenetically closely related suggesting that, being often in mixed infections, they probably had common ancestors. On the other hand, an additional conclusion can be drawn from the dendrogam in Figure 1f, that the unclassified tobamoviruses (in bold) did not constitute a homogenous cluster. In fact, they were irregularly dispatched through the investigated accessions.

Discussion

Positive selection and recombination are two major driving forces in virus evolution. Their study is of utmost importance. RNA viruses are renowned for their ability to evolve rapidly, a consequence of high mutation rates, rapid replication and large population sizes (Domingo and Holland, 1997). Moreover, there is mounting evidence for a role for recombination in shaping genetic diversity in some RNA viruses, particularly the retroviruses and a variety of positive-sense RNA viruses including potyviruses (Bousalem et al., 2003), luteoviruses (Gibbs, 1995), nepoviruses (Le Gall et al., 1995), closteroviruses (Vives et al., 2005), cucumoviruses and bromoviruses (Fraile et al., 1997; Bonnet et al., 2005) and most strikingly the caulimoviruses in which the rate of recombination per base exceeds that of mutation (Froissart et al., 2005). Likewise, the identification of defecting RNAs in many viruses (Cheng and Nagy, 2003) as well as rearrangements in the genome structure of more distantly related RNA viruses (Blok and Gibbs, 1995) also suggests a role for recombination. Despite some studies carried out by different virologists on recombination potentially occurring in the genome of numerous plant RNA viruses (Chare and Holmes, 2006), knowledge is still scare on evolution mechanisms in the new constituted family: *Virgaviridae*. Herein, molecular investigations in the whole genome of 168 accessions extracted from GenBank were conducted. Contrary to the statement of Adams et al (2009), the family *Virgaviridae* is not so homogenous. In this regard, inter-genera analyses resulted inconclusive. As a consequence, analyses were done within each genus. In this study, two methods were chosen (RECCO and RDP v.3.31β) based on the fact that they are appropriate for the mosaic structure of viruses as reported in previous works (Boulila, 2009a,b,c, 2010). Thus, the recombination frequency along the sequences of screened accessions of the *Virgaviridae* was dissected and basically revealed that: (i) recombination events were widespread in the genome of several members of the six genera composing this family, (ii) recombination events were more frequent in members of *Furovirus*, *Pecluvirus*, *Pomovirus*, and *Tobravirus* genera than those of *Hordeivirus* and *Tobravirus* genera, (iii) in majority, swapped fragments had a length exceeding three nucleotides reaching a size of as long as 952 residues (TMV.152 isolate), (iv) as assessed by RDP package, heredity material exchange arose between non homologous sequences like those of IPCV and PCV (*Pecluvirus* genus).

The determining of the selection pressures that have shaped genetic variation form a major part of many studies of molecular evolution. A common approach to this problem involves estimating the rates of nonsynonymous (d_N) and synonymous (d_S) substitutions. Kimura (1977) was the first to point out that their comparison can be a powerful tool for studying natural selection. Estimates of d_N significantly different from d_S, provide convincing evidence for nonneutral evolution. This approach is attractive as it does not make any assumptions regarding the demographic history of the population, unlike many neutrality tests, e.g. Tajima's test (1989) which compare estimates of effective population size obtained using different measures of genetic variation. In this study and according to Fisher's Exact Test of Neutrality, selective pressure investigations acting on protein expression of virus genes led to the identification of positive selection in two representatives of *Pomovirus* genus: the virus BSBV, and the virus PMTV. In addition, the tobamovirus ToMV and the unclassified tobamovirus MarMV were also under positive selection. Remarkably, these viruses evolved also under recombination mechanism. A synergism might be occurring. Such a synergism between recombination and natural selection may have played a major role in Darwinian molecular evolution.

In inferring evolutionary relationships of homologous sequences of members of the family *Virgaviridae*, it can be concluded that: (i) within segmented viruses, each delineated cluster corresponded to a divided genomic region, (ii) TMV isolates retrieved into three groupings, (iii) ToMV and TMV may have common ancestors, (iv) unclassified tombusvirus members did not constitute a phylogenetically homogenous group.

The accurate detection of recombination from DNA sequences becomes very relevant and indeed a number of methods have been developed for that purpose. Regarding RECCO, it has shown in this study that it is capable to detect recombinants in all the studied genera of the *Virgaviridae*. Also, it determined the frequency of recombination events through the aligned sequences. By contrast to RECCO, RDP package did not allow to reach such conclusions. In addition, the results obtained in both studies by RDP package were in some cases in line with those of RECCO and in others not so. In the author's opinion, the main advantage of RDP 3.31β algorithm was to indicate the potential donators of the genetic information to recombinants. Based on these observations, although RECCO is likely to give the most satisfactory results, both RECCO and RDP package should be considered as complementary methods, however.

Finally, in the author's knowledge, this is first report on *Virgaviridae* evolution in which molecular recombination structure and selective pressure as well as phylogenies were analyzed.

References

A. Fraile, J.L. Alonso-Prados, M.A. Aranda, J.J. Bernal, J.M. Malpica, F. Garcia-Arenal, *J. Virol.* 71, 934-940 (1997)

A.E. Greene, R.F. Allison, *Science* 263, 1423-1425 (1994)

B.G. Weiss, S. Schlesinger, *J. Virol.* 65, 4017-4025 (1991)

C.P. Cheng, P.D. Nagy, *J. Virol.* 77(22), 12033-12047 (2003)

D. Martin, E. Rybicki, RDP: detection of recombination amongst aligned sequences. *Bioinformatics* 16, 562-563 (2000)

D. Posada, K. Crandall, *Proc. Natl. Acad. Sci. USA* 98, 13757-13762 (2001)

D.P. Martin, C. Williamson, D. Posada, *Bioinformatics* 21, 260-262 (2005b)

D.P. Martin, D. Posada, K.A. Crandall, C. Williamson, *AIDS Res. H. Retrovir.* 21, 98-102 (2005a)

E. Domingo, J.J. Holland, *Annu. Rev. Microbiol.* 51, 151-178 (1997)

E.R. Chare, E.C. Holmes, *Arch. Virol.* 151, 933-946 (2006)

E.V. Koonin, V.V. Dolja, *Crit. Rev. Biochem. Mol. Biol.* 28, 375-430 (1993)

F. Tajima, *Genetics* 123, 585-595 (1989)

J. Blok, A.J. Gibbs, Cambr. Univ. Press. Cambridge, pp. 270-289 (1995)

J. Bonnet, A. Fraile, S. Sacristan, J.M. Malpica, F. Garcia-Arenal, *Virology* 332, 359-368 (2005)

J. Maydt, T. Lengauer, *Bioinformatics* 22(9), 1064-1071 (2006)

J.A. Hartigan, *Biometrics* 29, 53-63 (1973)

J.H. Strauss, E.G. Strauss, Evolution of RNA viruses. *Ann. Rev. Microbiol.* 42 (1988) 657-683.

J.M. Smith, *J. Mol. Evol.* 34, 126-129 (1992)

K.B. Pathak, P.D. Nagy, *Viruses* 1, 895-919 (2009)

M. Alejska, A. Kurzyniska-Kokorniak, M. Broda, R. Kierzek, M. Figlerowicz, *Acta Bioch. Polon.* 48, 391-407 (2001)

M. Baroth, M. Orlich, HJ. Thiel, P. Becher, *Virology* 278, 456-466(2000)

M. Boulila, *Biochem. Genet.* 48, 357-375 (2010)

M. Boulila, *Phytopathol. Mediterr.* 48, 411-421 (2009a)

M. Boulila, *Plant Mol. Biol. Rep.* 27, 189-198 (2009b)

M. Boulila, *Virus Genes* 38(3), 435-444 (2009c)

M. Bousalem, S. Dallot, S. Fuji, K.T. Natsuki, *Infect. Genet. Evol.* 3, 189-206 (2003)

M. Desport, M.E. Collins, J. Brownlie, *Virology* 246, 352-361 (1998)

M. Gibbs, Camb. Univ. Press Cambridge, pp. 351-368 (1995)

M. Hajjou, J.R. Hill, S.V. Subramaniam, J.Y. Hu, R. Raju, *J. Virol.* 70, 5153-5164 (1996)

M. Kimura, *Nature* 38, 515-524 (1977)

M. Nagai, Y. Sakoda, M. Mori, M. Hayashi, H. Kida, H. Akashi, *J. Gen. Virol.* 84Pt2 447-452 (2003)

M. Nei, T. Gojobori, *Mol. Biol. Evol.* 3(5), 418-426 (1986)

M. Padidam, S. Sawyer, C.M. Fauquet, *Virology* 265, 218-225 (1999)

M. Worobey, E.C. Holmes, *J. Gen. Virol.* 80, 2535-2543 (1999)

M.A. Ayllon, C. Lopez, J. Navas-Castillo, M. Mawassi, D.O. Dawson, J. Guerri, R. Flores, P. Moreno, *J. Gen. Virol.* 80, 817-821 (1999)

M.A. Larkin, G. Blackshileds, N.P. Brown, R. Chenna, P.A. McGettigan, H. McWilliam, F. Valentin, I.M. Wallace, A. Wilm, R. Lopez, J.D. Thompson, T.J. Gibson, D.G. Higgins, *Bioinformatics* 23(21), 2947-2948(2007)

M.C. Vives, L. Rubio, A. Sambade, T.E. Mirkov, P. Moreno, J. Guerri, *Virology* 331,232-237 (2005)

M.F. Boni, D. Posada, M.W. Feldman, *Genetics* 176, 1035-1047 (2007)

M.J. Adams, J.F. Antoniw, J.Kreuze, *Arch. Virol.* 154, 1967-1972 (2009)

M.J. Gibbs, J.S. Armstrong, A.J. Gibbs, *Bioinformatics* 16, 573-582 (2000)

M.M.C. Lai, *Microbiol. Rev.* 56, 61-79 (1992)

N. Galtier, M. Gouy, C. Gautier, *CABIOS* 12(6), 543-548 (1996)

OL. Le Gall, M. Lanneau, T. Candresse, J. Dunez, *J. Gen. Virol.* 76, 1279-1283 (1995)

R. Aaziz, M. Tepfer, *J. Gen. Virol.* 80, 1339-1346(1999)

R. Froissart, D. Rose, M. Uzest, L. Galibert, S. Blanc, Y. Michalakis, *PloS Biol.* 3,e89 (2005)
S. Kumar, M. Nei, J. Dudley, K. Tamura, *Brief. Bioinfor.* 9(4), 299-306 (2008)
S.Y. Morozov, A.G. Solovyer, *J. Gen.Virol.* 84, 1351-1366 (2003)
S.Y. Wang, J.C. Wu, T.Y. Chiang, Y.H. Huang, C.W. Su, I.J. Sheen, *J. Virol.* 81(9) 4438-4444(2007)
U. Melcher, *J. Gen. Virol.* 81, 257-266 (2000)
W.N. Fitch, *Syst. Zool.* 20, 406-416 (1971)
Y. Suzuki, *Gene* 365, 125-129 (2006)
Y. Suzuki, T. Gojobori, *Mol. Biol. Evol.* 16(10), 1315-1328 (1999)

In: Bioinformatics Research: New Developments
Editors: C. Battik and K. Belhassine
ISBN: 978-1-61942-363-3

Chapter VII

Molecular Dynamics Studies on the Structural Stability of Wild-Type Rabbit Prion Protein: Surface Electrostatic Charge Distributions

Jiapu Zhang *

Centre for Informatics and Applied Optimization and Graduate School of Sciences, Information Technology and Engineering, The University of Ballarat, MT Helen Campus, Australia

Abstract

Prion diseases cover a large range of neurodegenerative diseases in humans and animals, which are invariably fatal and highly infectious. By now there have not been some effective therapeutic approaches or medications to treat all prion diseases. Fortunately, numerous experimental experiences have showed that rabbits are resistant to infection from prion diseases isolated from other species, and recently the molecular structures of rabbit prion protein and its mutants were released into protein data bank. Prion diseases are "protein structural conformational" diseases. Thus, in order to reveal some secrets of prion diseases, it is amenable to study rabbits by techniques of the molecular structure and its dynamics. Wen et al. (PLoS One 5(10) e13273 (2010), Journal of Biological Chemistry 285(41) 31682-31693 (2010)) reported the surface of NMR $RaPrP^{C}$(124-228) molecular snapshot has a large land of continuous positive charge distribution, which contributes to the structural stability of rabbit prion protein. This is just the property at one snapshot. This paper confirms a large land of positive charge distribution has still been reserved during the long molecular dynamics of 30 nanoseconds in many environments, but the continuous of the land has not been always reserved yet. These results may be useful for the medicinal treatment of prion diseases.

* E-mail address: jiapu_zhang@hotmail.com, j.zhang@ballarat.edu.au.

Keywords: Prion diseases; Immunity; Rabbit prion protein; Molecular dynamics; Surface electrostatic charge distribution

Introduction

Prion diseases are invariably fatal and highly infectious neurodegenerative diseases affecting humans and animals. The neurodegenerative diseases such as Creutzfeldt-Jakob disease (CJD), variant Creutzfeldt-Jakob diseases (vCJD), iatrogenic CreutzfeldtJakob disease (iCJD), familial CreutzfeldtJakob disease (fCJD), sporadic CreutzfeldtJakob disease (sCJD), Gerstmann-Straussler-Scheinker syndrome (GSS), Fatal Familial Insomnia (FFI), KURU in humans, Scrapie in sheep, bovine spongiform encephalopathy (BSE or mad-cow disease) in cattle, chronic wasting disease (CWD) in white-tailed deer, elk, mule deer, moose, transmissible mink encephalopathy (TME) in mink, feline spongiform encephalopathy (FSE) in cat, exotic ungulate encephalopathy (EUE) in nyala, oryx, greater kudu, and spongiform encephalopathy (SE) in ostrich etc belong to prion diseases (en.wikipedia.org/wiki/Prion). By now there have not been some effective therapeutic approaches or medications to treat all these prion diseases.

Rabbits are one of the few mammalian species reported to be resistant to infection from prion diseases isolated from other species [1, 2, 3, 4, 5, 6, 7, 8, 9]. Thus, the paper studies rabbit prion protein in order to get some clues for the treatment of prion diseases.

Prion diseases are "protein structural conformational" diseases. The normal cellular prion protein (PrP^C) is rich in α-helices but the infectious prions (PrP^{Sc}) are rich in β-sheets amyloid fibrils. The conversion of PrP^C to PrP^{Sc} is believed to involve a conformational change from a predominantly α-helical protein (about 42% α-helix and 3% β-sheet) to a protein rich in β-sheets (about 30% α-helix and 43% β-sheet) [10]. Fortunately, the X-ray and NMR protein structures of rabbit prion protein and its S173N and I214V mutants were released into protein data bank (www.rcsb.org) (PDB ID codes 3O79, 2FJ3, 2JOH, 2JOM) recently. Hence, the conformational changes may be amenable to study by molecular dynamics (MD) techniques of these X-ray and NMR structures.

The author has studied the structural dynamics of rabbit prion protein and found that the structures were unfolded from α-helices into β-sheets under the conditions of 350 K or 450 K in low pH environment [11, 12, 13, 14, 15, 16]. However, under these conditions, the α-helices of human, mouse, dog and horse prion proteins were not found to unfold [11, 12, 13, 14, 15, 16]. The author concluded that some salt bridges (e.g. the one between residues 163 and 177) are clearly contributing to the structural conformational change [11, 12, 13, 14, 15, 16]. β2-α2 loop was reported to play an important role to stabilize the structural stability of rabbit and horse prion proteins [17, 18, 19, 20, 21, 22, 23, 24, 25, 26, 27, 28], and recently Wen et al. (2010) reported that rabbit prion protein has a unique distribution of large-continuous-positive -charge-surface electrostatic potential [25, 24, 21], which is with the highly ordered and well NMR-signal-recognized β2-α2 loop contributing to the structural stability of rabbit prion protein. The paper will study (by MD techniques) the dynamics of the surface electrostatic potential distributions of the 3D NMR structure of rabbit prion protein.

Surface Electrostatic Charge Distributions

To describe the potential energy associated with a charge distribution the concept of the electrostatic potential is introduced. The electrostatic potential at a given position is defined as the potential energy of a test particle divided by the charge of this object. When the charge distribution is over a particular area then the distribution is called as surface charge distribution. Electrostatic potential surfaces are valuable in computer-aided drug design because they help in optimization of electrostatic interactions between the protein and the ligand. These surfaces can be used to compare different inhibitors with substrates or transition states of the reaction.

First, we show the surface electrostatic charge distributions of 3D NMR structures of rabbit, horse, dog, human, mouse, rabbit S173N mutant and rabbit I214V mutant prion proteins (Figure 1).

We may see in Figure 1 that wild-type rabbits have a larger land of positive charge distributions than horses, dogs, humans and mice. Figure 1 just illuminates to us the surface charge distribution of one snapshot of rabbit prion protein. In this paper we will study the dynamics of the surface charge distributions during long MD simulations of 30 ns to confirm the above observation.

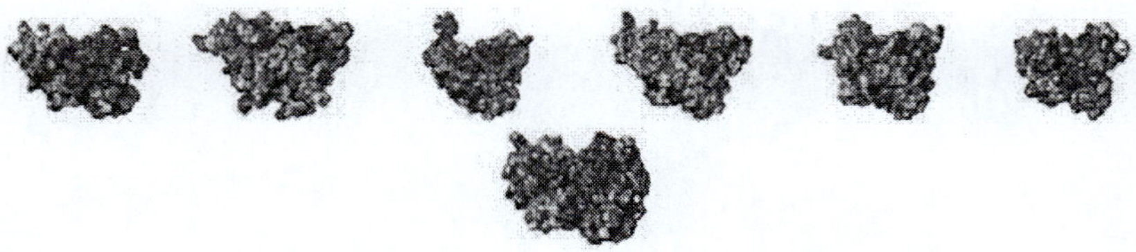

Figure 1. Distributions of the surface electrostatic potential of rabbit, horse, dog, human, mouse, rabbit S173N mutant, and rabbit I214V mutant prion protein NMR structures (PDB IDs 2FJ3 (Fig. 1a), 2KU4 (Fig. 1b), 1XYK (Fig. 1c), 1QLX (Fig. 1d), 1AG2 (Fig. 1e), 2JOH (Fig. 1f), 2JOM (Fig. 1g)). Blue is for positive charge whereas red is for negative charge. This Figure and all other Figures in the below were generated using MolMol (ftp://ftp.mol.biol.ethz.ch/software/MOLMOL/win/, http://www.csb.yale.edu/userguides/graphics/molmol/tutorial/example4.html).

Md Techniques

MD simulation materials are completely same as the ones of [11, 12, 13, 14] and the initial simulation structures were built on $RaPrP^C$(124-228) (PDB entry 2FJ3). MD simulation methods are completely same as the ones of [11, 12, 13, 14]. The next subsection illuminates the surface charge distributions of $RaPrP^C$ on the snapshots at 5 ns, 10 ns, 15 ns, 20 ns, 25 ns, 30 ns of 350 K and 450 K for different initial velocities seed 1 and seed 2, in neutral and low pH environments (Figures 2-9).

Surface Charge Distributions of RaPrP

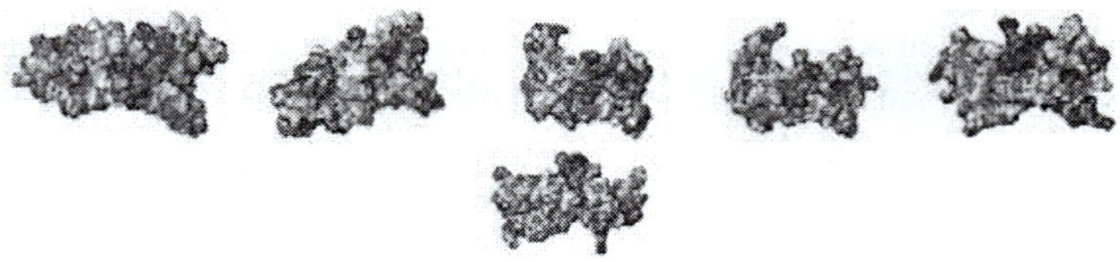

Figure 2. Distributions of the surface electrostatic potential of rabbit prion protein structure at 5 ns, 10 ns, 15 ns, 20 ns, 25 ns, 30 ns at 350 K, for seed 1 in low pH environment.

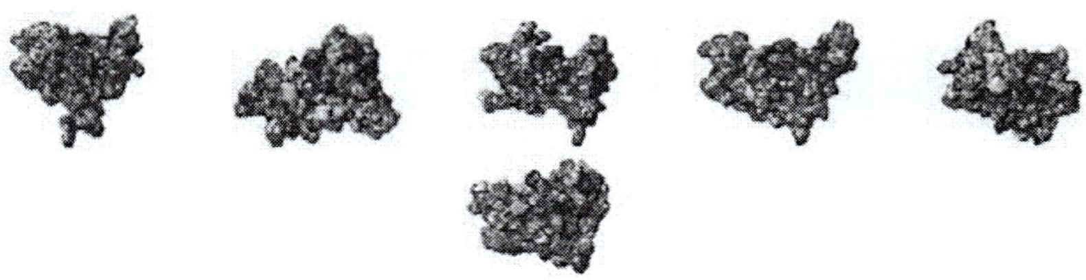

Figure 3. Distributions of the surface electrostatic potential of rabbit prion protein structure at 5 ns, 10 ns, 15 ns, 20 ns, 25 ns, 30 ns at 350 K, for seed 1 in neutral pH environment.

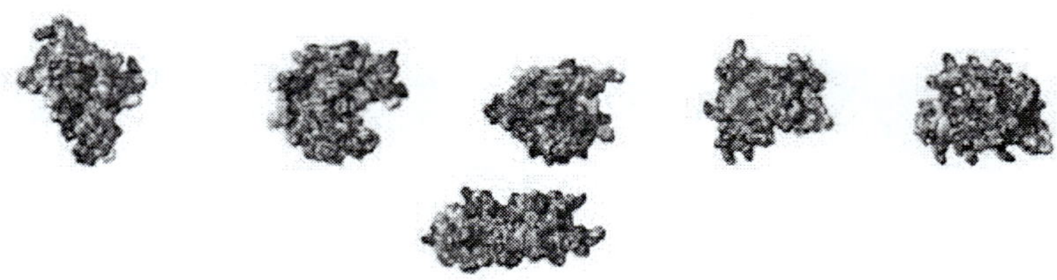

Figure 4. Distributions of the surface electrostatic potential of rabbit prion protein structure at 5 ns, 10 ns, 15 ns, 20 ns, 25 ns, 30 ns at 450 K, for seed 1 in low pH environment.

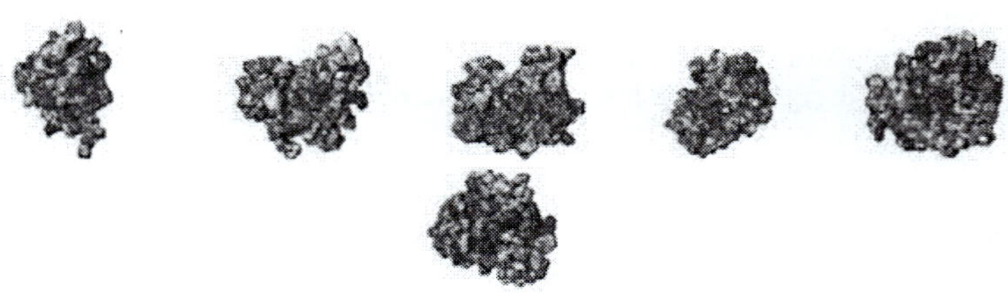

Figure 5. Distributions of the surface electrostatic potential of rabbit prion protein structure at 5 ns, 10 ns, 15 ns, 20 ns, 25 ns, 30 ns at 450 K, for seed 1 in neutral pH environment.

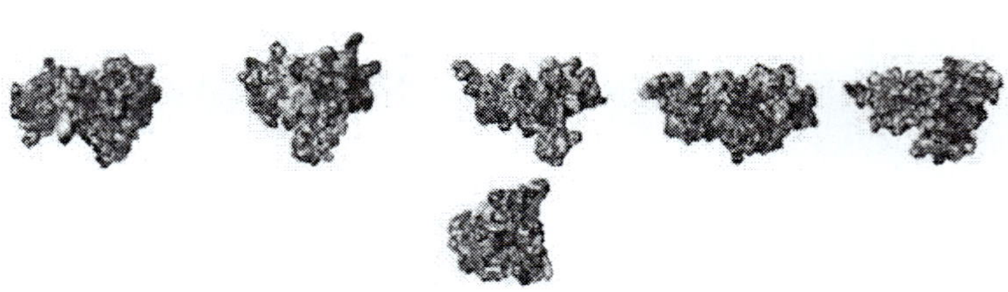

Figure 6. Distributions of the surface electrostatic potential of rabbit prion protein structure at 5 ns, 10 ns, 15 ns, 20 ns, 25 ns, 30 ns at 350 K, for seed 2 in low pH environment.

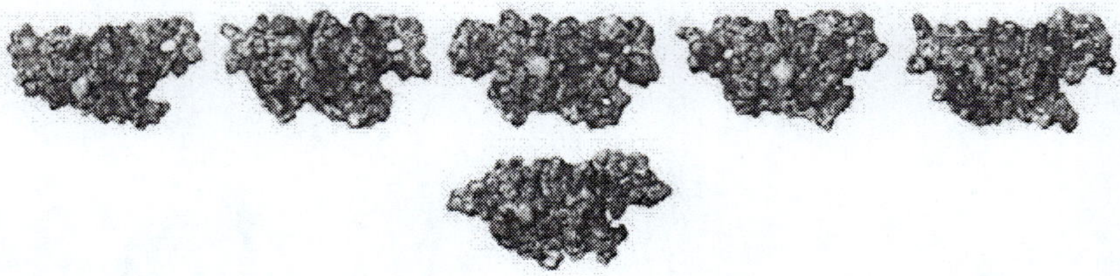

Figure 7. Distributions of the surface electrostatic potential of rabbit prion protein structure at 5 ns, 10 ns, 15 ns, 20 ns, 25 ns, 30 ns at 350 K, for seed 2 in neutral pH environment.

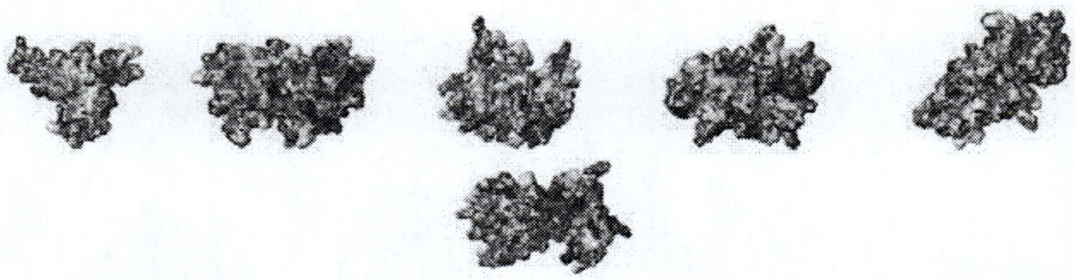

Figure 8. Distributions of the surface electrostatic potential of rabbit prion protein structure at 5 ns, 10 ns, 15 ns, 20 ns, 25 ns, 30 ns at 450 K, for seed 2 in low pH environment.

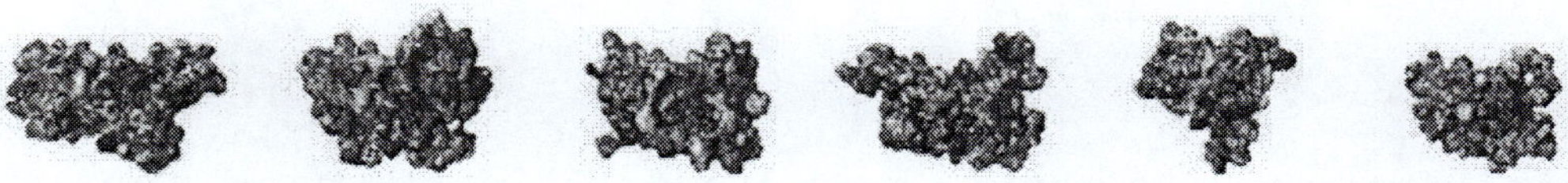

Figure 9. Distributions of the surface electrostatic potential of rabbit prion protein structure at 5 ns, 10 ns, 15 ns, 20 ns, 25 ns, 30 ns at 450 K, for seed 2 in neutral pH environment.

Analyses of Surface Charge Distributions of RaPrP

We may see in Figures 2-9 that the continuation of blue or red colors for low pH value is less for neutral pH value; this might imply to us RaPrP has less structural stability in low pH environment. Similarly, we may list the snapshots of surface charge distributions of HoPrP^C(119–231) (PDB entry 2KU4), DoPrP^C(121–231) (PDB entry 1XYK), HuPrP^C(125-228) (PDB entry 1QLX) and MoPrP^C(124-226) (PDB entry 1AG2) during the long MD simulations of 30 ns. However, we cannot find their clear differences from RaPrP^C(124-228). In the below, we only analyze the surface charge distributions of special residues ARG163 (Figure 10) and ASP177 (Figure 11) and of the β2-α2 loop RaPrP(163-171) (Figure 12).

By observations from Figure 12 we cannot see much difference among these snapshots at the 30th nanosecond. Figures 10-11 illuminate to us that the NMR snapshot has large land of uniform distributions around the the residues R163 and D177 respectively, but we cannot find the real difference among all the snapshots at the 30th nanosecond.

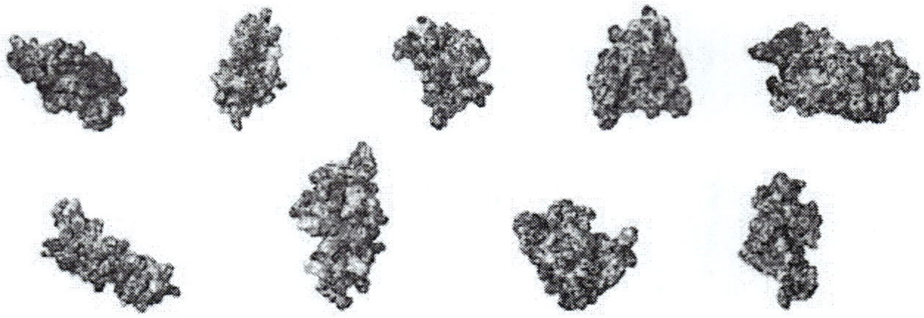

Figure 10. Distributions of the surface charges around residue R163 (labeled into the yellow circle) of RaPrP^{C}(124-228) at snapshots of 2JF3.pdb (Fig. 10a), the 30th nanosecond of low pH - seed 1 - 350 K, low pH - seed 2 - 350 K, neutral pH - seed 1 - 350 K, neutral pH - seed 2 - 350 K, low pH - seed 1 - 450 K, low pH - seed 2 - 450 K, neutral pH - seed 1 - 450 K, neutral pH - seed 2 - 450 K.

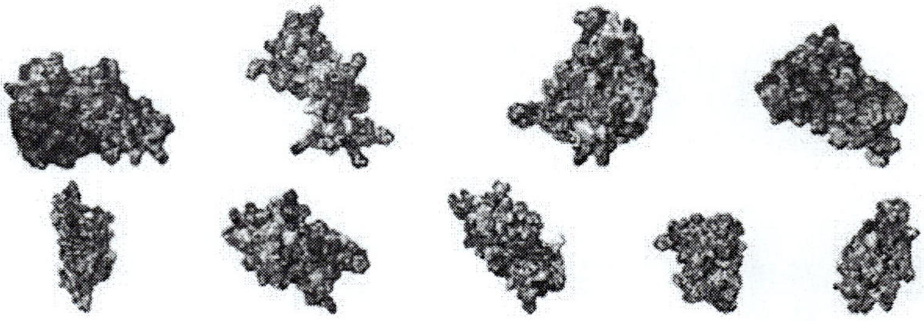

Figure 11. Distributions of the surface charges around residue D177 (labeled into the yellow circle) of RaPrP^{C}(124-228) at snapshots of 2JF3.pdb (Fig. 11a), the 30th nanosecond of low pH - seed 1 - 350 K, low pH - seed 2 - 350 K, neutral pH - seed 1 - 350 K, neutral pH - seed 2 - 350 K, low pH - seed 1 - 450 K, low pH - seed 2 - 450 K, neutral pH - seed 1 - 450 K, neutral pH - seed 2 - 450 K.

Figure 12. Distributions of the surface charges around the β2-α2 loop RaPrP(163-171) (labeled into the yellow circle) at snapshots of 2JF3.pdb, the 30th nanosecond of low pH - seed 1 - 350 K, low pH - seed 2 - 350 K, neutral pH - seed 1 - 350 K, neutral pH - seed 2 - 350 K, low pH - seed 1 - 450 K, low pH - seed 2 - 450 K, neutral pH - seed 1 - 450 K, neutral pH - seed 2 - 450 K.

Conclusion

The surface of NMR RaPrP^C(is still reserved during the long MD simulations of 30 ns under many different conditions, this paper presents numerous snapshots of graphs for NMR RaPrP(124-228). These graphs should have a value acted as a rich database to study the 'unique distribution of surface electrostatic potential' of NMR RaPrP^C(124-228) molecule. 124-228) molecule has a large land of continuous positive charge distribution (Figure 13). This agrees with the observation of [25, 24, 21]. To check whether this continuous positive charge distribution.

Generally, we find that, in neutral pH environment, rabbit prion protein has a large land of positive charge distribution (Figures 1a, 1g, 3, 5, 7, 9, 10a, 11a); this agrees with the finding of Wen et al. [25, 24, 21]. In low pH environment, the large continuous and disappears (Figures 2, 4, 6, 8) and the structural stability of rabbit prion protein is broken (from α-helices into β-sheets). In conclusion, the surface electrostatic potential charge distribution, more or less, contributes to the structural stability of rabbit prion protein and contributes to the study of the species barrier of rabbit prion protein to prion diseases.

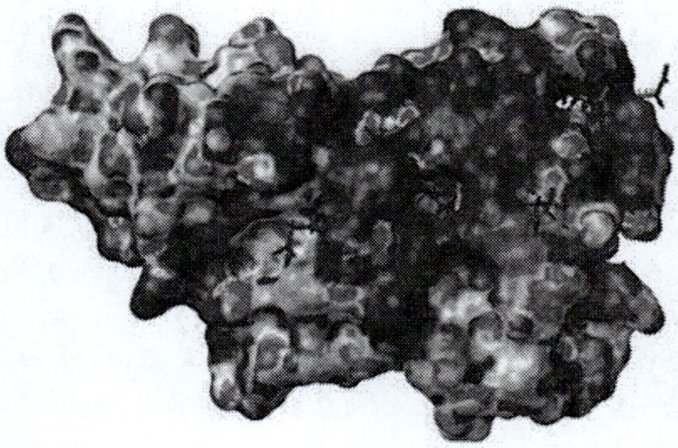

Figure 13. The continuous distribution of positive surface charges of NMR RaPrP^C(124-228) (PDB entry 2FJ3).

Acknowledgments

This research has been supported by a Victorian Life Sciences Computation Initiative (VLSCI) Grant Number VR0063 on its Peak Computing Facility at the University of Melbourne, an initiative of the Victorian Government.

References

[1] I. Vorberg, H. G. Martin, P. Eberhard, A. P. Suzette, *J. Virol.* 77(3), 2003 (2003).

[2] B. Loftus, M. Rogers, *Gene* 184(2), 215 (1997).

[3] R. Rubenstein, R. J. Kascsak, M. Papini, R. Kascsak, R. I. Carp, G. LaFauci, R. Meloen, J. Langeveld, *J. Neurovirol.* 5(4), 401 (1999).

[4] R. J. Kascsak, R. Fersko, D. Pulgiano, R. Rubenstein, R. I. Carp, *Immunol. Invest.* 26(1-2), 259 (1997).
[5] L. Souan, Y. Tal, Y. Felling, I. R. Cohen, A. Taraboulos, F. Mor, *Eur. J. Immunol.* 31(8), 2338 (2001).
[6] V. Bringue, J. L. Vilotte, H. Laude, *Vet Res.* 39(4), 47 (2008)
[7] R. M. Nisbet, C. F. Harrison, V. A. Lawson, C. L. Masters, R. Cappai, A. F. Hill, *J. Virol.* 84(13), 6678 (2010).
[8] P. Fernandez-Funez, Y. Zhang, S. Casas-Tinto, X. Xiao, W. Q. Zou, D. E. Rincon-Limas, *J. Biol. Chem.* 285(47), 36897 (2010).
[9] R. A. Moore, I. Vorberg, S. A. Priola S.A., *Arch. Virol. Suppl.* 19, 187 (2005).
[10] J. S. Griffith, *Nature* 215, 1043 (1967).
[11] J. P. Zhang, *J. Theor. Biol.* 264(1), 119 (2011).
[12] J. P. Zhang, *J. Theor. Biol.* 269(1), 88 (2011).
[13] J. P. Zhang, D. D. W. Liu, *J. Biomol. Struct. Dyn.* 28(6), 861 (2011).
[14] J. P. Zhang, *J. Biomol. Struct. Dyn.* 29(2), 369 (2011).
[15] J. P. Zhang, The nature of the infectious agents: PrP models of resistant species to prion diseases (dog, rabbit and horses). In: *Prions and Prion Diseases: New Developments (J.M. Verdier Eds.)*, NOVA Publishers, 2011.
[16] J. P. Zhang, Computational studies of the structural stability of rabbit prion protein compared with human and mouse prion proteins. In: *Neurodegenerative Diseases / Book 2 (R.C.C. Chang and J.Y.S. Ho Eds.)*, INTECH Open Access Publisher, 2011, ISBN 979-953-307-672-9.
[17] D. R. Perez, F. F. Damberger, K. Wuthrich K., *J. Mol. Biol.* 400(2), 121 (2010).
[18] P. Fernandez-Funez, Y. Zhang, J. Sanchez-Garcia, K. Jensen, W. Q. Zou, D. E. Rincon-Limas D.E., *Commun. Integr. Biol.* 4(3), 262 (2011).
[19] M. Q. Khan, B. Sweeting, V. K. Mulligan, P. E. Arslan, N. R. Cashman, E. F. Pai, A. Chakrabartty A., *Proc. Natl. Acad. Sci. USA* 107(46), 19808 (2010).
[20] M. Q. Khan, *Misfolding of particular PrP species and susceptibility to prion infection.* University of Toronto, Master Degree Thesis (2010)
[21] D. H. Lin, Y. Wen, *Scientia Sinica Chimica* 41(4), 683 (2011).
[22] B. Sweeting, M. Q. Khan, A. Chakrabartty, E. F. Pai E.F., *Biochem. Cell Biol.* 88(2), 195 (2010).
[23] B. Sweeting, E. Brown, A. Chakrabartty, E. F. Pai, *Canadian Light Source* 28 (2009): http://www.lightsource.ca/science/pdf/activity report 2009/28.pdf
[24] Y. Wen, J. Li, M. Q. Xiong, Y. Peng, W. M. Yao, J. Hong, D. H. Lin, *PLoS One* 5(10), e13273 (2010).
[25] Y. Wen, J. Li, W. M. Yao, M. Q. Xiong, J. Hong, Y. Peng, G. F. Xiao, D. H. Lin, *J. Bio. Chem.* 285(41), 31682 (2010).
[26] I. Biljan, G. Ilc, G. Giachin, A. Raspadori, I. Zhukov, J. Plavec, G. Legname, *J. Mol. Biol.* doi:10.1016/j.jmb.2011.07.067 (2011)
[27] P. Christen, C. F. Damberger, K. R. Perez, S. Hornemann, K. Wuthrich, Cellular prion protein conformation and function, and Temperature-dependent conformational exchange in the cellular form of prion proteins. Journal: To be *Published* (2011).
[28] K. Kaneko, L. Zulianello, M. Scott, C. M. Cooper, A. C. Wallace, T. L. James, F. E. Cohen, S. B. Prusiner S.B., *Proc. Natl. Acad. Sci. USA* 94(19), 10069 (1997).

In: Bioinformatics Research: New Developments
Editors: C. Battik and K. Belhassine
ISBN: 978-1-61942-363-3

Chapter VIII

Bioinformatics Analysis of Gene Networks Involved in Genomic Stability and Cancer

Marialva Sinigaglia [5]***, José L. Rybarczyk Filho*** [3]***,***
Rodrigo J. S. Dalmolin[4]***, José C. F. Moreira*** [4]***,***
Rita M. C. de Almeida[3]***, Mauro A. A. Castro***[2]***,***
José C. M. Mombach[2] ***and Giovani R. Librelotto*** [1]
[1]Departmento de Informática, Universidade Federal de Santa Maria, Santa Maria, Brazil
[2]Departmento de Física, Universidade Federal de Santa Maria, Santa Maria, Brazil
[3]Instituto de Física, Universidade Federal do Rio Grande do Sul, Caixa Postal, Brazil
[4]Departamento de Bioquímica, Universidade Federal do Rio Grande do Sul, Rua Ramiro, Brazil
[5]Departamento de Genética, Universidade Federal do Rio Grande do Sul, Caixa Postal, Brazil

Abstract

Chromosomal abnormalities, somatic mutations and epigenetic alterations are among the major cellular changes associated withthe impairment of genome maintenance mechanisms (GMM) in solid tumors. Together with apoptosis, the GMM gene network is essentially characterized by the four major DNA repair mechanisms: nucleotide excision repair (NER), base excision repair (BER), mismatch repair (MMR) and recombinational repair (RER)—homologous recombinational repair (HRR) and nonhomologous endjoining (NHEJ). The GMM gene network is highly interconnected. Its dysfunction contributes substantially to cancer genome instability, which hasimportant clinical implications, since it imposes technical problems for tumor biomarker development, such as the low recurrence of somatic mutations causally related to cancer and the extensive tumor heterogeneity. Using statistical, systems biology and computational approaches, in this chapter we consider different types of molecular changes that can imbalance the genome of cancer cells together with current methods used to study high-throughput data in the context of apoptosis and genome stability systems. In the introduction we discuss the molecular features that are commonly attributed to solid tumors, the theories that describe the tumor progression and the resulting DNA damage accumulation patterns. In section two we review some computational methods used to characterize gene expression networks, some available internet resources that can be used to study genome instability and the possible strategies to assess tumor biomarker networks. Finally, in section three, we present several applications illustrating the use of the computational methods previously defined, from the analysis of cancer-related gene networks using microarray data and gene knockout, to the analysis of tissue homeostasis and cell growth control in the perspective of possible evolutionary scenarios that gave rise to genome maintenance mechanisms in multicellular organisms.

1. Introduction

The analysis of genomic studies involves the development of new bioinformatics methods to deal with the available large-scale molecular data. The production of transcriptome, proteome and interactome (i.e., estimated amount of mRNAs, proteins and molecular interactions) are examples of approaches that require advanced computational strategies to infer any significant association according to the hypothesis being tested. One of these strategies is the search for sets of biomarkers able to discriminate, for example, a pathological condition from their normal state, which could help to stratify patients in cohort studies using data from functional genomics.

The large amount of molecular information and professionals working on this topic, and the indisputable relevance of the area, has led to the formation of public repositories, supported or managed by international organizations such as EMBL, NIH/NCBI and KEGG. Databases of protein interactions have different origins: they may be the result of experiments of high performance (from high-throughput technology) or may be the result of research studies focused on a few genes or proteins, compiled over decades of study and classified according to some criterion, as,for example, participating in certain biochemical functions, acting on the same substrate, expressed in one specific cell or tissue, etc.

The size and complexity of biological data collected from such technologies requires searching for innovative methods to analyze the resulting datasets (Golemis and Adams 2005). Is it possible to construct valid principles and methods that could be used to explore the wealth of molecular information available? What computational tools need to be designed for this task? The technological counterpart of these issues is also important. If new bioinformatics tools were validated for genome association studies, it would surely be of fundamental importance for the discovery of new drugs, new treatments and new diagnostic tests. However, it is still a challenge to obtain correlations and propose or validate new methods from this wealth of information.

One of these challenges in the post-genomics is the understanding of the structure and behavior of molecular networks that control cell functioning (Barabási and Oltvai 2004). The scenario in question is an intricate network of interactions between genes, proteins and other biochemical components of cells. The complexity of such a network can be perceived from metabolic maps and graphs of protein-protein interactions, as illustrated in Figure 1.

An example that illustrates the difficulty in applying genomic technologies in medicine comes from the current solid tumors staging system - TNM system (Greene *et al.* 2002). The TNM classification is used to record malignant tumor progression and to stratify patients by severity, which provides essential information for choosing initial treatment, prognosis and patient follow-up. According to the current staging system, only germ line tumors are staged with the aid of molecular markers (i.e., TNM+S system).

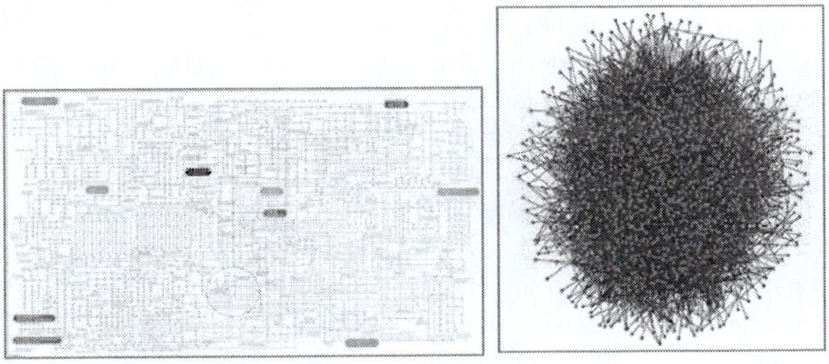

(A) (B)

Figure 1. Metabolic maps and protein networks. (A) KEGG reference pathways (http://www.genome.jp/). (B) Graph of protein-protein interactions of 1892 proteins of Homo sapiens – (Castro et al. 2009 – Supplements).

In other solid tumors, however, the staging is done exclusively by anatomical criteria, despite the many tools available for molecular analysis of the genome, proteome and transcriptome of biopsies of malignant tissue (Ogino and Goel 2008). Reinforcing this difficulty is the fact that no biomarker described in the post-genomic era has been formally used (Kulasingam and Diamandis 2008), even those identified in carcinogenesis study models, as *APC*, *TP53* and *KRAS* genes in colorectal cancer.

A plausible explanation for this apparent discrepancy is the difficulty in achieving the basic requirements for aggregating potential biomarkers to the formal staging systems. An ideal biomarker should be sensitive and specific, should have a good effective cost and should be robust against operational variability (Ludwig and Weinstein 2005). The great tumor heterogeneity is a major barrier to accomplish these requirements (Ransohoff 2005;Kulasingam and Diamandis 2008). Recently, several studies suggest a new paradigm for the development of molecular biomarkers (Bild *et al.* 2006; Potti *et al.* 2006; Chen *et al.* 2007). With the use of genomic technologies, instead of searching for a biomarker in particular, these studies seek groups or patterns of biomarkers, whose signature may be able to discriminate phenotypes and outcomes of a given disease. Therefore, as biomarkers alone does not seem to be effective to translate tumor heterogeneity into clinically relevant information, the robustness of the molecular classification of solid tumors could be improved by identifying networks of tumor biomarkers, which would be used to propose new tumor stratifications.

1.1. Biomarkers, Bioinformatics and Solid Tumors

Few types of cancer account for most of the fatal cancer cases. For example, in Brazil more than 90.0% of all cancer deaths in both sexes is associated with the development of solid tumors. Carcinomas of lung, stomach and breast cancer account for almost 30% of all cases (Atlas de Mortalidade por Câncer 2009). Considering only lung cancer, it is a universe of approximately 1.3 million deaths a year worldwide, which makes this solid tumor among the five leading causes of death in developing countries, ahead of hypertension or cardiac complications, or even traffic accidents—data from WHO (World Health Organization 2009).

Unlike hematological malignancies, solid neoplasm form tumor masses that are generally restricted to tissue barriers for long periods during the early stages of cancer progression, when it acquires invasive characteristics that can erupt from the origin compartment and colonize new sites (Jones *et al.* 2008).

One aspect extensively reported is that solid tumors acquire a very large number of sporadic molecular changes throughout its progression, taking the natural history of a given disease to different paths, with many potential outcomes for the same tumor type (Lengauer *et al.* 1998; Flyger *et al.* 1999; Osterheld *et al.* 1999; Berczi *et al.* 2002; Hoglund *et al.* 2002; Jin *et al.* 2002; Albertson *et al.* 2003; Jass *et al.* 2004; Hoglund *et al.* 2005; Heng *et al.* 2006; Shibata 2006; Shen *et al.* 2007). As a result, in advanced stages these cancers exhibit extensive intratumoral heterogeneity, with formation of multiple aberrant clones (Harada *et al.* 1998; Glockner *et al.* 2002; Losi *et al.* 2005; Backvall *et al.* 2005).

Chromosomal abnormalities, somatic mutations and epigenetic changes are among the major cellular changes that reflect the increasing heterogeneity of solid tumors. Chromosomal abnormalities in solid tumors are characterized by a high degree of numerical and structural aberrations (Duesberg *et al.* 2005) -- numerical aberrations correspond to changes in the standard number of chromosomes of a species, while structural aberrations correspond to anomalies in the structure of chromosomes. The numerical aberrations may result from changes in both the haploid set of chromosomes (euplody) and in parts of this set (aneuploidy), generating chromosomal unbalances. According to the international system for cytogenetic nomenclature of the human species (ISCN 1995), more than 30 terms are used to record cytogenetic aberrations. Translocations, deletions and additions of material of unknown origin are among the most frequent types of structural aberrations found in karyotypes of solid tumors, while the spectrum of numerical aberrations varies from 20 to 300 chromosomes, with high occurrence of aneuploid karyotypes around the diploidy (Mitelman Database 2009). Due to this large degree of abnormality, recurrent aberrations -- with some clinical significance, are not easily found in solid tumors (Albertson *et al.* 2003), but recent studies show that measures of diversity (or entropy) can allow the use of karyotypes for prognosis (Hoglund *et al.* 2005; Merlo *et al.* 2006; Maley *et al.* 2006).

Just as chromosomal aberrations, somatic mutations are present in large quantities in solid tumors (Venkatesan *et al.* 2006). A recent study pointed out that from 18,191 genes investigated in colon and breast cancer, 9.4% had at least one silent mutation, and most of these changes (92.7%) corresponds to point mutations (Wood *et al.* 2007). This point mutations can be passengers or causal. It is understood that *causal* refers to those mutations whose underlying role is directly related to tumor progression, conferring some adaptive advantage to the neoplastic cell population (Futreal *et al.* 2004).

According to estimates from the Cancer Gene Census (Futreal 2009), 347 human genes have somatic mutations designated as causally related to cancer development. Passenger mutations, on the other hand, are present in low frequency and do not confer adaptive advantages, but can be distinguished from changes expected for a normal cell (e.g., nucleotide polymorphism) (Greenman *et al.* 2007).

However, the distinction between causal and passenger mutations can be quite superficial when considering the spectrum of changes across a cancer sample. For example, in a global gene mutation estimate, Wood and coworkers (Wood *et al.* 2007) have found low levels of agreement of causative mutations in 11 samples of DNA sequenced colon cancers (the same happened with breast cancer), a difference that the authors suggest as the bases of the large variability of solid tumors. Moreover, epigenetic modifications can alter the functioning of genes associated with tumor progression without the need for changes in nucleotide sequence. For example, changes in DNA methylation pattern in regions with rich CG dinucleotide content (CpG islands) is one of best characterized epigenetic modifications in cancer. These CpG islands are located preferentially in gene regulatory regions and may obstruct the coupling of transcription factors when covalently modified by the addition of methyl groups (-CH3) (Feinberg *et al.* 2006). The hypermethylation of tumor suppressor genes seems to be a common event in colon cancer (Shen *et al.* 2007).

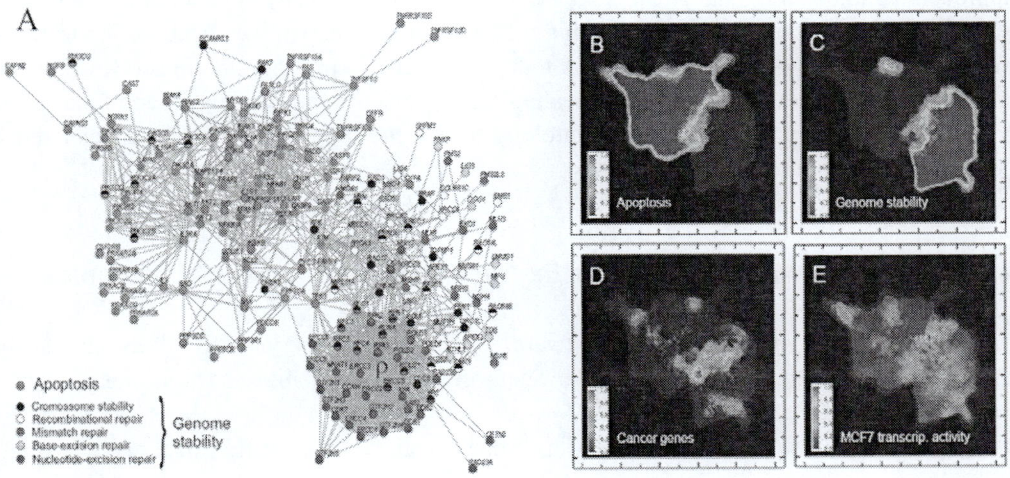

Figure 2. Analysis of human genome maintenance mechanisms. (A) Graph of protein-protein interactions of genome stability pathways and apoptosis. (B-E) Functional mapping on the graph of interactions. B: apoptosis gene area; C: genome stability gene area; D: mutations causally associated with cancer development; E: transcriptional activity of MCF7 breast cancer cell line. Adapted with permission from NAR (Oxford University Press): Castro *et al.*, Nucleic Acids Research, 35: 1859-1867, 2007; Reviewed at: Castro *et al.*, Bioinformatics, 25:1468–1469, 2009.

Therefore, changes in genome stabilizing systems are directly involved with the increased heterogeneity of solid tumors. These systems are concentrated in three different and partial overlapped cellular processes: (i) mechanisms of DNA repair (Hoeijmakers 2001), (ii) mechanisms of segregation and stabilization of chromosomes (Jefford and Irminger-Finger 2006) and (iii) mechanisms of programmed cell death (apoptosis) (Zhivotovsky and Kroemer 2004).

A network model of the genome maintenance mechanisms (GMM) is depicted in Figure 2A (Castro *et al.* 2007). This network model of molecular interactions is an example of a bioinformatics platform that can be used to transfer molecular data between various functional databases (Castro *et al.* 2008). This task can be performed by using softwares specially designed for this purpose, as the software ViaComplex, which plots different types of information - as gene expression data - onto the network topology (Castro *et al.* 2009) (Figure 2B-E). In the next section we review the software tools, databases and statistical methods used in our systems biology approach to study and characterize the normal and altered functional states of the GMM.

2. Tools and Methods

An important goal of bioinformatics is the development of models that integrate the different types of 'omics' data available. As an alternative to reach this goal we developed the software ViaComplex. There are several free software tools available to manage gene network interactions (interatomics), however none of them allows the visualization of gene expression data (transcriptomics) on gene networks in a three-dimensional landscape approach coupled with statistical analysis of gene and/or pathway expression levels as ViaComplex does. Below we describe this software. In this section we also describe the most important source of gene expression data today, the Gene Expresion Omnibus (GEO), where we can find the data used in papers dealing with transcriptomics analysis and finally, we present the Ontocancro database and ontology that is the only tool available specialized in genome maintenance mechanisms.

2.1. Landscape Analysis of Gene Expression Networks with Viacomplex

Viacomplex is an open-source software. It was developed in FORTRAN and Dislin scientific library (http://www.dislin.de/). The software is available at (http://lief.if.ufrgs.br /pub/biosoftwares/viacomplex/).

The main feature of this program is associated with the distribution of a given quantity (quantitative or qualitative data) onto gene/protein interaction networks. It is an application for gene expression data analysis that builds landscape graphics of genome maintenance gene networks. It provides tools that may help potential users to investigate different high-throughput experiments in the perspective of genome maintenance mechanisms (GMM), as exemplified at Figure 2. The software can project gene activity data onto this internal model, or import customized models of gene/protein interaction networks. It can be used to produce publication quality images where data is visualized as functional landscapes projected onto the human GMM map.

The internal GMM network model is built by direct (physical) and/or indirect (functional) associations according to STRING Database (Snel *et al.* 2000). In customized options, it is possible to use other networks from different organisms. It can map, for example, cancer genes (Figure 2D), transcriptional activity (Figure 2E), cell and tissue lethality (Castro *et al.* 2008).

As additional resources, the software can calculate the degree of functional overlap among all nodes of a given network according to the approach described by Ravasz et. al (2002). Likewise, it provides a statistical module to assess Groups of Functionally Associated Genes (GFAG) based on the concept of the information theory. ViaComplex considers two quantities to estimate gene expression alterations in GFAGs: the distribution of the signal in the pathway (i.e. gene expression diversity) and the intensity of the signal, where the problem of multiple comparisons is controlled by the False Discover Rate (FDR) approach (Benjamini *et al.* 1995), suitable for genome-scale studies.

2.2. The Gene Expression Omnibus (Geo)

The Gene Expression Omnibus (GEO) is a public repository developed and maintained by the National Center for Biotechnology Information (NCBI) to host high-throughput gene expression datasubmitted by the scientific community. These data are generated by different technologies such as microarray, serial analysis of gene expression (SAGE), and next-generation sequencing. The GEO database standardizes and storages this avalanche of information and provides, for users, tools to search, analyze, visualize, and perform the download of these huge volumes of data. The search can be performed in several ways, including basic and advanced searches as well as using nucleotide sequences to begin the query. Once the DataSet of interestis identified, it's possible to finds genes differentially expressed between experimental subgroups using the data analysis tool "compare 2 sets of samples" or visualize cluster heat map images and select regions of interest for further study using "cluster heatmaps". Currently, the GEO database has received thousands of data submission comprising more than 452.000 samples and 16 billion individual abundance measurements for over 500 organisms (Barret et al, 2009). GEO is accessible at http://www.ncbi.nlm.nih.gov/geo/.

2.3. The Ontocancro Database and Ontology

Disruption of the mechanisms that regulate cell-cycle checkpoints, DNA repair, and apoptosis results in genomic instability and the development of cancer. The description of the complex network of these pathways requires news tools to integrate large quantities of experimental data in the design of biological information systems. In this context this chapter presents Ontocancro: an extensible ontology and an architecture designed to facilitate the integration of data originating from different public databases in a single- and well-documented relational database. Ontocancro is an ontology stored in a knowledge database designed to be a source of information to integrate transcriptomics and interatomics data involved in gene pathways of Genome Maintenance Mechanisms (GMM). In pratice this data integration is difficult since biological data are disseminated in various databases, lack a unified vocabulary, and need constant update. Ontocancro aims to facilitate this task through integration of information on genome stability pathways from curated databases as NCI-Nature, BioCarta, KEGG, Reactome, Prosite, GO, and also from databases on physical interactions among proteins as STRING. Ontocancro is accessible at: www.ontocancro.org.

Due to a lack of consensus in the gene set definition of each different pathway in the databases and in the scientific literature, we developed the Ontocancro pathways that include all documented genes (from databases and literature) listed in these pathways (Castro *et al.* 2008). Ontocancro pathways are meant to be a comprehensive gene lists for each pathway. The expanded Ontocancro pathways include, additionally, the genes that interact physically with these genes (interactions determined from the STRING database at 0.9 confidence level).

Ontologies and Topic Maps - Everyday thousands of new biological information resources are linked to the web through databases. This way the web is growing very fast, making search tasks more and more difficult. To solve the problem several initiatives were undertaken and a new area of research and development emerged: the one called Semantic Web. When the authors refer to the Semantic Web they are thinking about a network of concepts. Each concept has a group of related resources and can be related to other concepts, though it can use this concept network to navigate among web resources or simply among information resources. From the undertaken initiatives two have become standards: Ontologies and Topic Maps. Ontology is a way of describing a shared common understanding, about the kind of objects and relationships which are being talked about, so that communication can happen between people and application systems (Guarino, 1998). In other words, it is the terminology of a domain (it defines the universe of discourse). The component that represents and maintains these structured collections is the ontology. An ontology is a logical theory to relate the intended meaning of a formal vocabulary, i.e., it is a binding with a particular conceptualization of a world. The ontology includes structures that allow manipulating terms in a more efficient way; it is useful to the human understanding and validation mechanisms operating on inter-agents communication (Bittner *et al*, 2004). The importance of its use is the ability to represent hierarchies of object classes (taxonomies) and their relationships. In the same area, different ontology definitions and classifications can be found. In Artificial Intelligence, Guarino defines ontology as a logical theory accounting for the intended meaning of a formal vocabulary, i.e. its ontological commitment to a particular conceptualization of the world (Bittner *et al.* 2004). In the area of bioinformatics, ontology is defined as a set of concepts and terms, that can be used to describe some area of knowledge, or construct a representation of the knowledge (Backer *et al.* 1999). Ontologies are theories of content about the object types, object properties, and relationships between objects that are possible in a domain of specific knowledge. An ontology consists of a set of axioms which place constraints on sets of individuals ("classes") and the types of relationships permitted among them. These axioms provide semantics by allowing systems to infer additional information based on the data explicitly provided. The data described by an ontology is interpreted as a set of "individuals" and a set of "property assertions" which relate these individuals to each other. The ontology's development will continue to provide the construction mechanism of the semantic part of Semantic Web. The development of these mechanisms depends of languages that express the information in a way that machines can understand. The challenge is to provide a language that will enable the manipulation, in an efficient way, of data and rules for queries about these data, and that will allow existing rules in any knowledge representation system to be exported to the Web. The ontology's development will have to represent an important part of the effort in the development of any application in the future. That way, the development of environments to the construction and manipulation of ontologies is crucial and important. Such environment must be composed by an ontology deposit that can be manipulated by developers, users, and application programs, allowing the navigation, search and terms reuse. When new terms are added to the ontology, the environment must verify the deposit consistency and act accordingly.

Topic Maps are abstract structures that can encode knowledge and connect this encoded knowledge to relevant information resources (Park and Hunting, 2003). Topic Maps allow a domain knowledge representation in semantic networks, composed of topics and associations. Topic Maps is a formalism to represent knowledge about the structure of a set of information resources and to organize it in topics. A topic, in its most generic sense, can be anything. A person, an entity, a concept, really anything regardless of whether it exists or has any other specific characteristics. It constitutes the basis for the topic maps creation. A topic can have one or more occurrences. An occurrence represents the information that is specified as relevant to a given topic. One or more addressable information resources of a given topic, constitutes the occurrence. Occurrences and topics exist on two different layers (domains), but they are connected. Occurrences establish the routes from the topics to the information resources, providing the reason why that route exists from a topic to an information object. These topics also have associations that represent and define relationships between them. As described in ISO/IEC 13250, a topic association specifies a relationship among specific topics (e.g. that "Pathways contains Genes" or "Databases are source of Pathways"). A topic association is a link between topics, each of which plays a role as a member of that association. This is important because a relationship that holds between topics is probably interesting even without the given context that the topic and its associations were created for (e.g. Pathways are available on Biological Databases, that a set of pathways available on biological databases relation is valid in most contexts). Information about the topics can be inferred by examining the associations and occurrences linked to the topic. A collection of these topics and associations is called a topic map. Topic Maps can be seen as a description of what there is about a certain domain, by formally declaring topics, and by linking the relevant parts of the information set to the appropriate topics (Wache *et al.* 2001). A topic map expresses someone's opinion about what the topics are, and which parts of the information set are relevant to which topics. Charles Goldfarb (Park and Hunting, 2003) usually compares Topic Maps to a GPS (Global Positioning System) applied to the information universe. Talking about Topic Maps is talking about knowledge structures. Topic Maps are the basis for knowledge representation and knowledge management. Enabling to create a virtual map of information, the information resources are kept in their original form and so they are not changed. Then, the same information resource can be used in different ways, for different topic maps. As it is possible and easy to change the map itself, information reuse is achieved. Topic Map architecture was also designed to allow merging between topic maps without requiring the merged topic maps to be copied or modified, enabling an incremental growth. In fact topic maps would indeed gain effectiveness and interoperability either through explicit formalization of ontologies specifically built and dedicated for topic map control, or through declaration of commitment to pre-defined ontologies, not specifically designed for that use. Topic Maps take the main concepts described in the information resources and relates them together independently of what is said about them in the information being indexed. This means managing the meaning of the information, rather than just the information because topic maps take a step back from the details and focusing on the forest rather than the trees. XML Topic Maps (XTM) 1.0 specification Park and Hunting, 2003 was developed by topicmaps.org which is an independent consortium of parties developing the applicability of the topic map paradigm (ISO/IEC 13250:2000) to the World Wide Web by leveraging the XML family of specifications (Bradley, 2002).

2.3.1. Ontocancro Pathways

Genomic DNA is under constant attack from both endogenous and exogenous sources of DNA damaging agents. During the process of evolution, organisms have acquired a series of mechanisms to repair DNA damage, thus assuring genome stability and faithful transmission of genetic information. Defects in Genome Maintenance Mechanisms (GMM) contribute significantly to the onset of cancer (Hoeijmakers, 2001). Ontocancro is an Ontology that integrates the information of the most important Human pathways involved in GMM: apoptosis, nucleotide excision repair (NER), base excision repair (BER), homologous recombinational repair (HR), nonhomologous end joining (NHEJ), mismatch repair (MMR), cell cycle (CC) and chromosome stability (CS). Each pathway contains a comprehensive list of genes collected from different databases such as NCI-Nature, BioCarta, KEGG, Reactome, Prosite, Gene Ontology (GO) and others. The participating genes were manually curated and include information about interactions, enabling a global view of the protein networks that link all of these processes. It is possible to download the information about the pathways in Ontocancro as shown in Figure 3.

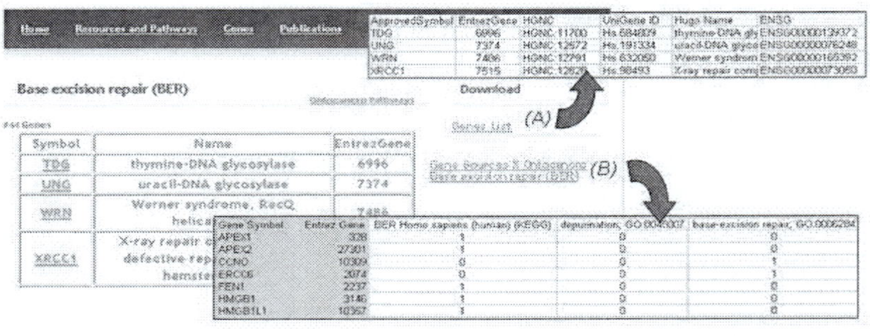

Figure 3. (A) 'Genes List' is list of genes belonging to a pathway with additional information like approved Symbol, Hugo Name, Entrez Gene, Ensembl ID (ENSG), etc.; (B) The 'Gene Source X Ontocancro' pathway is a table with information about genes provenance. A number 1 in the table means that the gene is present in a given pathway. This way the User can choose which genes he wants to keep or remove the pathway according to its source.

Ontocancro includes 120 pathways related to GMM and a total of 1428 genes at the moment. Information about human DNA repair pathways was obtained from the paper of Wood *et al.* (2001), KEGG and GO and the chromosome stability pathway was included in accord to the work of Vogelstein (Rajagopalan et. al., 2003).

The complex response of cells against DNA injury triggers the intricate interactions between DNA repair, cell-cycle progression and apoptosis. Considering that defective apoptosis represents a major causative factor in the development and progression of cancer and that the molecular pathways of cellular death still need better elucidation, we made a broad search for genes involved in human programmed cell death in the literature and in bioinformatics databases. To generate an Ontocancro apoptosis pathway, we started from the apoptosis pathway described in the KEGG database containing about 100 genes. We enriched this list of genes with other genes mentioned to be involved in apoptosis found in databases as the GO, NCI-Nature, Reactome, and BioCarta. We also included genes that contain death domains involved in apoptosis from the Prosite database. From this manual search we generated a set of 491 genes involved in cell death that we call Ontocancro apoptosis. To make this search even broader, given the lack of information regarding this pathway, we used these 491 genes as a 'seed pathway' to search for additional genes that interact physically with them and that could also be involved in apoptosis based in the 'guilty by association' inference, since interacting genes can have similar function. For that we used the STRING database with input options 'experiments' and 90% confidence level. From this last analysis we found 464 additional genes that may be functionally involved in apoptosis and we called this pathway expanded apoptosis with a total of 955 genes.

2.3.2. Ontocancro's Architecture

Daily, a lot of data is stored into biological databases and its organization requires an integrated view of their heterogeneous information systems. In this situation, there is a need for an approach that extracts the information from their data sources and fuses it in a semantic network. Usually, this is achieved either by extracting data and loading it into a central repository that does the integration before analysis, or by merging the information extracted separately from each resource into a central knowledge base. This is the Ontocancro's goal.

A major difficulty to the improved integration of pathway databases (such as KEGG, NCI-Nature, GO, Reactome, Prosite, and BioCarta) and gene annotation databases (like UniGene and HGNC) is the absence of up to date and well-maintained mapping between corresponding entries. Fortunately, all this databases make available our content in XML files for download. XML files contain mainly the data about genes and pathways.

However this XML files are structured according to different schemas (basically one for each database). To achieve this aim, we propose the architecture described in figure 4. We developed a parser to extract the data from biological databases and store them into several XML files (Bradley, 2002). These generated files are stored into a XML database management system, called eXist.

In this context, we have applied the ontology paradigm to add a knowledge layer over this information system. Metamorphosis (Librelotto *et al.* 2008) links the information resources to a single ontology according to the Topic Maps format. We make Topic Maps a key component of the new generation of Web-aware knowledge management solutions. In this way, Metamorphosis let us achieve the semantic interoperability among heterogeneous information systems because the relevant data is extracted and stored in a topic map, according to the desired information specified on an ontology. This ontology is designed to facilitate the integration of data originating from different public information resources.

Metamorphosis (Librelotto *et al.* 2008) links the information resources (public biological databases) into an integrated view in a single- and well-documented relational MySQL database. This MySQL database contains the generated Topic Map and allows its processing by PHP scripts running over an Apache web server. This web server publishes the Ontocancro web page. Thus the navigation over the topic map is led by a semantic network and provides a homogeneous view over the resources. Ontocancro users can view and search information provided for these pathway and genes. Its homepage is shown in figure 5.

The Ontocancro page has a search engine. The results of a search are ranked so that terms most closely matching a query are returned first. Individual words and combinations of words are scored according to the field in which they occur and their frequency within Ontocancro. If a pathway is selected, all its genes are listed. Each gene name has a link to its own page. This feature is displayed in figure 6.

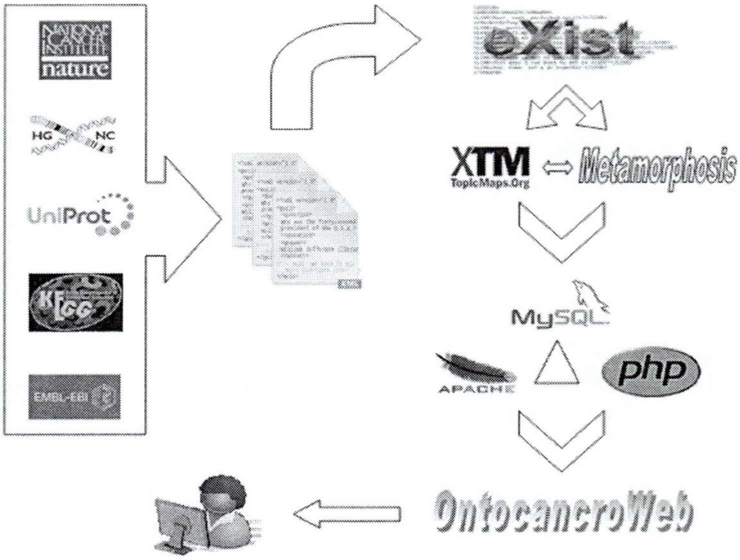

Figure 4. Ontocancro's architecture.

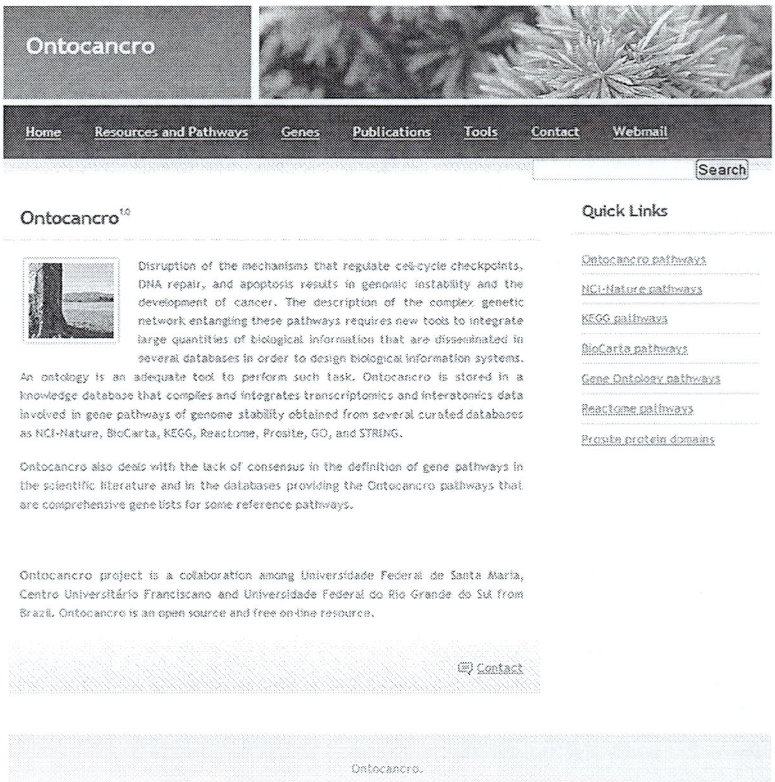

Figure 5. Ontocancro homepage.

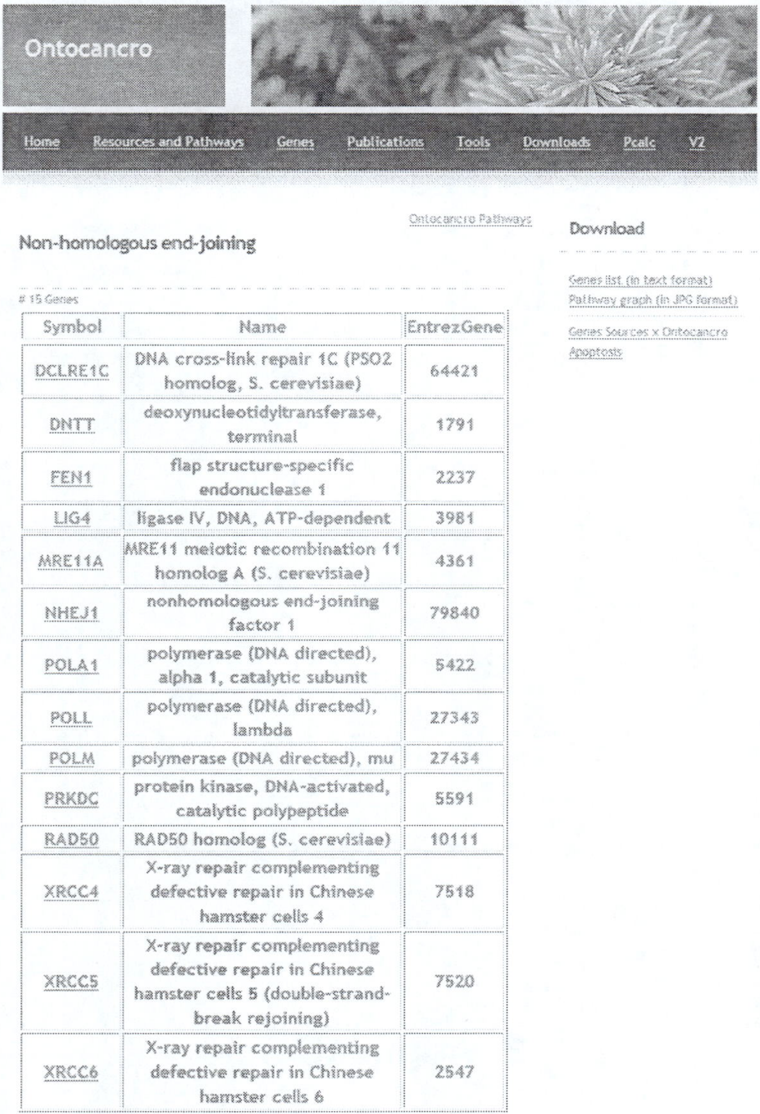
Ontocancro

Home Resources and Pathways Genes Publications Tools Downloads Pcalc V2

Ontocancro Pathways

Non-homologous end-joining

15 Genes

Symbol	Name	EntrezGene
DCLRE1C	DNA cross-link repair 1C (PSO2 homolog, S. cerevisiae)	64421
DNTT	deoxynucleotidyltransferase, terminal	1791
FEN1	flap structure-specific endonuclease 1	2237
LIG4	ligase IV, DNA, ATP-dependent	3981
MRE11A	MRE11 meiotic recombination 11 homolog A (S. cerevisiae)	4361
NHEJ1	nonhomologous end-joining factor 1	79840
POLA1	polymerase (DNA directed), alpha 1, catalytic subunit	5422
POLL	polymerase (DNA directed), lambda	27343
POLM	polymerase (DNA directed), mu	27434
PRKDC	protein kinase, DNA-activated, catalytic polypeptide	5591
RAD50	RAD50 homolog (S. cerevisiae)	10111
XRCC4	X-ray repair complementing defective repair in Chinese hamster cells 4	7518
XRCC5	X-ray repair complementing defective repair in Chinese hamster cells 5 (double-strand-break rejoining)	7520
XRCC6	X-ray repair complementing defective repair in Chinese hamster cells 6	2547

Download

Genes list (in text format)
Pathway graph (in JPG format)

Genes Sources x Ontocancro
Apoptosis

Figure 6. Ontocancro ‘Non-homologous end-joining’ pathway page.

When a gene is selected, Ontocancro’s search engine presents a tabbed view of the genes information (including approved symbol provided by HGNC, EntrezGene, and Affymetrix HG-U133 microarray platforms). A table composed of gene interacting partners is shown after the annotations of the selected gene. This page can be seen in figure 7.

All data provided by Ontocancro can be queried remotely, both for genes and pathway information. These web services are fully integrated, so that the filtering options and datasets available (one for each pathway) are fully synchronized between the browseble and web service interfaces. The web services have been designed for ease use, providing results displayed in TXT format. Outputs in XML and RDF will be available soon.

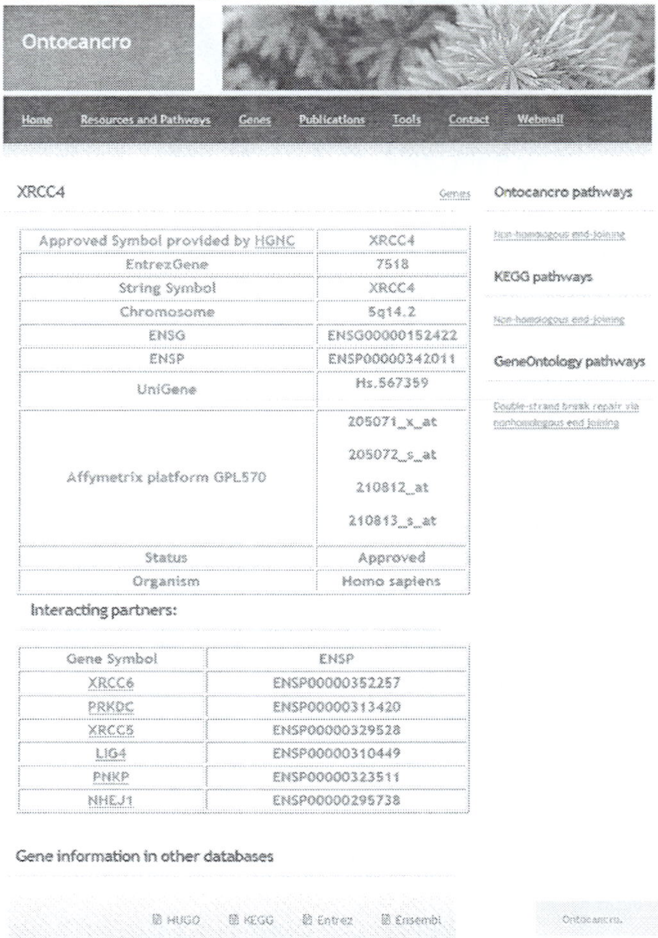

Figure. 7. *XRCC4* gene page.

3. Applications

Using the tools presented in the section Tools and Methods, we present two different studies involving the network of genome maintenance mechanisms in cancer. In the first study we compared the expression of the GMM in normal and cancer tissues using an alternative method of assessing gene expression, Serial Analysis of Gene Expression (SAGE) from the Cancer Genome Anatomy Project (CGAP) (http://cgap.nci.nih.gov/SAGE). In the second study we have built an evolutionary scenario for the GMM.

3.1. GMM Transcription Activity in Cancer

We analyzed both, the amount of transcripts production and its diversity in ten gene pathways obtained from SAGE analysis from the CGAP database. The pathways were chosen due to either their involvement in GMMand/or, as a control group, due to their essential life supporting activities (ribosome, ATP synthase, electron transport chain, and glycolysis). The tumor types were selected such that they present a library of normal cells, to be used as control. The details of the statistical analysis can be found in Castro *et al.*, (2007), here we only refer to the main results and conclusions of the paper.

We have compared the gene expression activity and diversity of each cancer SAGE library or exam from several tissue types with its respective normal case in the database. We have found a general increase in gene expression activity and diversity. In Figure 8we compare all gene expression pathways in relation to the normal case, we present the average values of the relative activity and diversity for each gene network. As we can observe, NER and apoptosis present the lowest relative activity (Figure 8A) and highest relative diversity (Figure 8B), supporting that the low level of gene expression is associated with changes in gene activity in this repair pathway.

Figure 8 shows that both apoptosis and NER networks are simultaneously affected, and so a causal correlation is plausible considering that both networks are entangled, see Figure 2(A). This result was surprising since in cancer cells programmed cell death mechanisms are in general structurally impaired (Zhivotovsky andKroemer, 2004), what is coherent with the observed gene expression profile of apoptosis.

However, NER is in general structurally intact in sporadic solid tumors, as no somatic mutations in NER genes have been reported to be causally implicated in oncogenesis (Futreal *et al.*, 2004). The observed transcript profile then suggests that NER transactivation dependent functions are affected in cancer cells.It is then possible that conditions that disable apoptosis, probably due to mutations in apoptotic genes, also affects NER transactivation dependent functions via *p53*. NER mal-functioning could then account for random point mutations scattered throughout the cancer cell genome, hence characterizing an acquired "mutator phenotype".

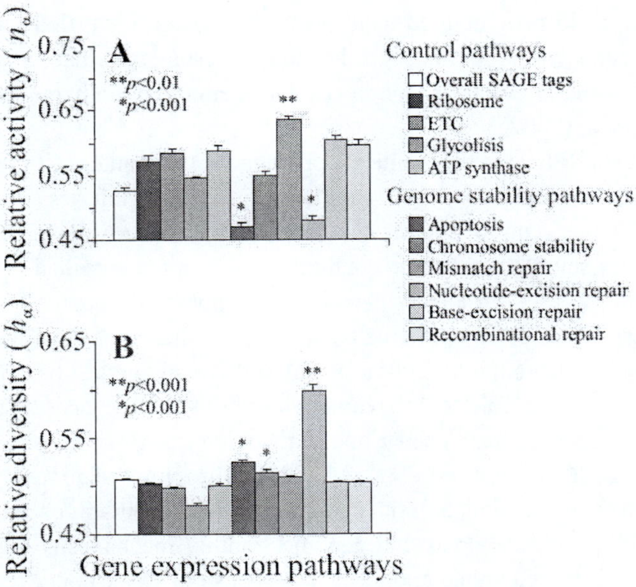

Figure 8. Comparison of relative activity and diversity for 10 different pathways in cancer. Adapted with permission from NAR (Oxford University Press): Castro et al., Nucleic Acids Research, 35: 1859-1867, 2007.

3.2. Evolutionary Aspects of the GMM

The earliest evidence of cellular life on earth comes from 3,500 Myr ago. There are lots of information gaps about those forms of life. In spite of being largely accepted that life must have originated in aquatic environment, it remains unclear the exact conditions that have provided its appearance. The biochemical strategy that those ancient organisms used to extract energy from their surroundings also remains controversial. Among the scarce information about the first cells it is known that they probably had a cellular membrane, were prokaryotes, and had nucleic material (Line, 2002; Whitfield, 2004). If an organism possesses a genome, it will need a mechanism to stabilize it. Thus, it is reasonable to think that genome maintenance mechanisms evolved early in evolutionary history. Reinforcing that, DNA repair systems are extensively recognized in prokaryotes suggesting the presence of these mechanisms in prokarya's Last Common Ancestor (LCA) (Helling,1968; Setlow and Carrier, 1964; Wildenberg and Meselson, 1975; Willetts and Clark, 1969). Similarly to the genome stability system, other biochemical mechanisms evolved through the unicellular era to increase cellular fitness therefore increasing organism fitness.

Around 1,200 Myr ago, a new organizational status of life appears: the multicellularity. After more than 2 billion years of cellular evolution, this new condition of life emerges (Carroll, 2001). One of the characteristic of multicellularity is that the individual cellular fate is not the protagonist of fitness, whereas the pluricellular organism is. To allow a friendly and cooperative coexistence among numerous cells, which may or may not present the same biological function in a tissue, former molecular mechanisms developed in the unicellular era needed to be adapted to this new reality. In this sense, the development of programmed cell death systems certainly gave an important improvement in individual cellular fate control for the benefit of a higher organizational level of life: the multicellular organism (Huettenbrenner *et al.*, 2003). Even if programmed cell death has been identified in prokaryotes and unicellular eukaryotes, a large body of information suggests that complex biochemical systems of cellular death, such as apoptosis, have emerged with multicellular organisms [reviewed at Ameisen, (2002)].

Apoptosis is a sophisticated biochemical apparatus that guarantees (i) the substitution of old and dysfunctional cells damaged by, for example, environmental injuries and (ii) the elimination of useless cells no longer requested in the developmental process. A dysfunctional apoptosis has been associated with several pathogenic conditions such as uncontrolled cell growth typically found in neoplastic diseases. In cancer perspective, suppressed apoptosis is a requirement to enhance cell fitness. Over proliferative cells can surpass the organism fitness altering body's homeostasis (Crespi and Summers, 2005; Danial and Korsmeyer, 2004; Hipfner and Cohen, 2004). Apoptosis has been related to genome instability in the sense that mutated clones carrying aberrant genetic content need a malfunctioning apoptosis to survive (Castro *et al.*, 2006; Yan *et al.*, 2006; Zhivotovsky *et al.*, 2004). The close relationship among genome-maintenance mechanisms and apoptotic components has been largely described including proteins that work in both apoptosis and genome stability mechanisms (e.g. *p53*), and DNA repair proteins, which can stimulate apoptosis in response to DNA injury (e.g. *PARP1*) (Alano *et al.*, 2004; Castro *et al.*, 2007; Sengupta and Harris 2005).

A network representation including apoptosis and genome-stability systems is shown in Figure 2A, where is noticable the entangled functioning of both systems and their coordinate operation is crucial to maintain cellular homeostasis and exercises a central role in cancer constriction, as it has been mentioned above. Taking into account the different evolutionary origins of each system, how can we explain the intimate relationship involving both systems? And, when do they start working together? To find an answer to those questions, Castro and colleagues constructed an orthology map across a species tree for the human apoptosis and genome-stability gene/protein-association network (Castro *et al.*, 2008). The work has been based in finding the LCA of each gene/protein depicted in Figure 2A. A species tree involving 35 eukaryotes with fully sequenced genomes was utilized to suggest the evolutionary root of each 180 genes comprising apoptosis and genome-stability systems. Figure 2B and C give us a topological representation of human apoptosis and genome-stability gene network obtained using the software ViaComplex (Castro *et al.*, 2009) showing the apoptosis prevalence area and genome-stability prevalence area, respectively. From now on, we will use landscape representation to show and discuss that network properties and Figure 2B and C can be used as a guide to locate apoptosis and genome-stability respective areas in the network topology.

According to the authors, while the majority of genome-stability orthologs were present in eukaryotic LCA, apoptosis orthologs are poorly represented in the base of species tree, suggesting the former arrives later in eukaryotic evolutionary history. Focusing in apoptosis, three major events were described: (i) the root of few apoptotic components in the eukarya origin (such as Cytocrome c); (ii) network enrichment with apoptotic core components of the intrinsic pathway (such as *BCL2* and the initiator caspase-9) in the metazoan origin; and (iii) network enrichment with apoptotic core components of the extrinsic pathway (such as several TNF superfamily members) in the origin of vertebrates (Figure9A). Figure 9B shows the cross-talk between apoptosis and genome-stability is the most derived portion of the network and has been populated along the evolutionary process. In contrast, genome-stability concentrates the ancestry of the network. It suggests genome-stability components start to work as a system in the eukaryotes origin or before that (Castro *et al.*, 2008). Since genome-stability system evolves early in evolution and has been found across a species tree - from unicellular eukaryotes to mammals - it is reasonable to think this biochemical machinery is crucial to eukaryotic cell homeostasis maintenance.

One of the ways used to evaluate the essentiality of a network is to evaluate the essentiality of its genes, *i.e.* whether a gene is or not required for the organism survival. Using the orthology map of human genome maintenance mechanisms, data from other organisms than *Homo sapiens* can be used to project functional information obtained from those species in human network as exemplified in Fig 10A.

In this figure, yeast and mouse were used to project essentiality information onto human apoptosis and genome-stability network. Yeast data was extracted from Saccharomyces Genome Database (http://www.yeast genome.org/) (Christie *et al.*, 2004) and mouse data was obtained from Mouse Genome Database (http://www.informatics.jax.org) (Bult *et al.*, 2008). Taking into account that yeast is a unicellular organism, knock-out orthologs which confer lethality were defined as essential at the cellular level. In the same way, mouse knock-out orthologs which confer lethality before zygote implantation were defined as essential at the cellular level. As we can see in the Figure 11A, the information concerning cellular essentiality matches the genome-stability area. This can be explained for the genome-stability system appeared before multicellularity and any alteration in its components can be lethal to the cell. Apoptosis, as discussed before, evolved after multicellularity appearance and impairment in its components can alter organism development and/or homeostasis even though it does not harm the cell.

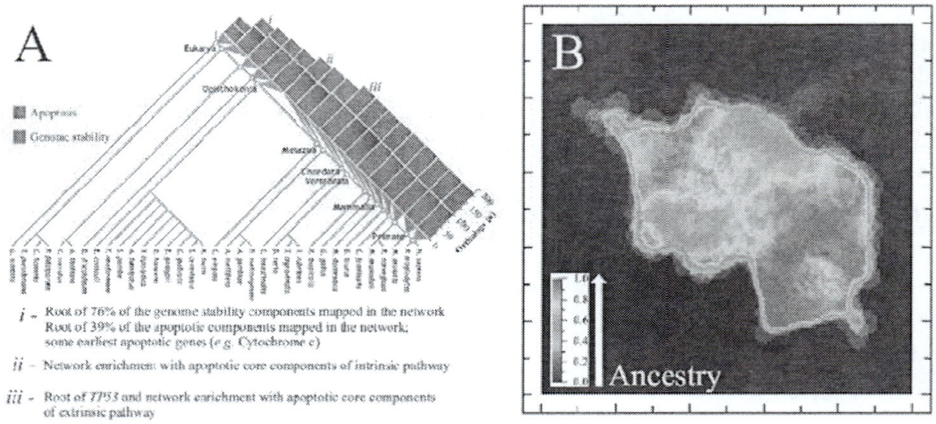

Figure 9. Evolutionary scenario of apoptosis and genome-stability gene network. A: Species tree used to infer evolutionary origin of apoptosis and genome-stability components. The histograms show the distribution of 180 human orthologs according to the roots inferred in the eukaryote species tree. B: Landscape representation apoptosis and genome-stability gene network. Adapted from Castro et al., Nucleic Acids Research, 36: 6269-6283, 2008.

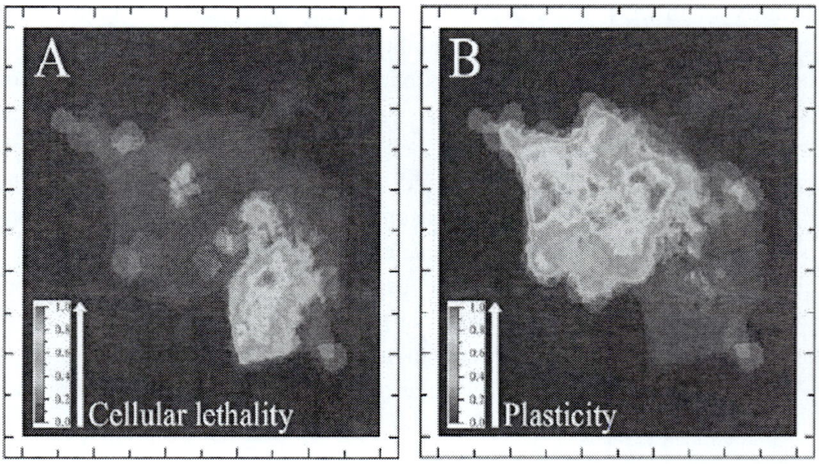

Figure 10. Network plasticity. A: Knock-out orthologs which confer lethality in *Saccharomyces cerevisiae* and knock-out orthologs which confer lethality before zygote implantation in *Mus musculus* were defined as essential at the cellular level. B: Plasticity of each network component has been defined according diversity and abundance of respective KOG.

Other interesting measure to evaluate evolutionary history of a gene network is to analyze its ortholog distribution in a species tree. Each gene of a given network has a set of orthologs that can be defined as an ortholog group (OG) found along a species tree. There are several databases that organize and classify OGs, such as the String database (http://string.embl.de/) (Jensen *et al.*, 2009), Keeg Database (http://www.genome.jp/kegg/) (Aoki-Kinoshita and Kanehisa, 2007), and the Inparanoid database (http://inparanoid.sbc.su.se/cgi-bin/index.cgi)

(Remm *et al.*, 2001). Orthologous genes not necessarily need to be equally distributed along a species tree, as long as an OG may have more genes associated with some species than others.

Using Shannon Information Theory (Shannon and Weaver, 1949) we can evaluate the distribution of orthologs of a given OG in a given species tree using a diversity measure (see mathematical definitions in the previous study in this section).

If a given orthologous group has only one ortholog in every organism of the species tree, this means that its diversity will be maximum (H_α=1.0). In contrast, if all orthologs of a given GO are present in only one organism the OG diversity will be minimum (H_α=0.0). A complementary quantity of ortholog distribution is obtained by calculating the ratio between the number of orthologous genes in a given OG and the number of organisms that have genes belonging to that OG (the abundance). If an OG has only one ortholog per organism (D_α=1.0) it indicates poor representation in any particular species. In contrast, KOGs with D_α>>>1.0 indicates many variants per organisms. Analyzing diversity and abundance together, an OG presenting maximum diversity and minimum abundance has orthologous genes in all species in a given species tree and only one ortholog per species. This ortholog had a constrained evolution, it was not deleted or duplicated despite the fact it probably had alterations in nucleotide sequence along its evolutionary history which was not sufficient to lose the orthology information. Taking into account that genetic plasticity may be understood as the ability of a functional gene network to tolerate changes in its components, diversity and abundance can be used collectively as a plasticity measure, considering that more plastic genes belong to OG with high diversity and low abundance and less plastic genes belong to OG with low diversity and high abundance. Figure 10 B shows the plasticity analysis of human apoptosis and genome-stability gene network using information from the tool 'Eukaryotic Clusters of Orthologous Groups' (KOGs) from the String database. As we can see, apoptosis concentrates the plasticity of the network while genome-stability presents low plasticity. These data add to information shown before: genome stability is the ancestral portion of the network, its orthologs were present in the origin of eukaryotes, and its components are essential to cellular homeostasis. Accordingly, while apoptosis is in the middle of its evolutionary process, genome-stability may have reached its evolutionary plateau. In the cancer perspective, we do not expect to find deletion in genome-stability components since it will probably make cells unviable and the neoplasm will not evolve. Reinforcing this idea, it is known that somatic mutations in nucleotide excision repair genes and base excision repair genes have never been described in cancer cells (Futreal *et al.*, 2004, Mombach *et al.*, 2008). Germline mutations have been described in those repair mechanisms; however mutations conferring nullity (i.e. gene mutations which lead to a condition equivalent to gene knock-out) are totally absent in homozygosis in those very rare syndromes (Castro *et al.*, 2008,Cleaver *et al.*, 1999).

According to the scenario presented here, apoptosis appearance seems to have been related to a developmental control as a requirement to a correct ontogeny in multicellular organisms. Since the production of cells in excess is a hallmark of metazoan development, apoptosis must act to eliminate the no longer requested cells during later stages of development. Therefore to maintain the relative number of cells of different types and to achieve proper organ function (Meier *et al.*, 2000). In addition, apoptosis linkage to genome-stability mechanisms has permitted the clean elimination of damaged cells which would inevitably die, avoiding intracellular material leakage to extracellular environment. This linkage among both systems most likely becomes useful in the cancer constraining role of apoptosis. The subsequent appearance of extrinsic pathway components at the origin of vertebrata represents an increment in complexity. The apoptosis extrinsic pathway is characterized by the ligation of cell surface receptors via specific death ligands to trigger caspase-8. The majority of these ligands belong to the tumor necrosis factor (TNF) superfamily, which includes FAS (*FASLG*) and TRAIL (*TNFSF10*). They bind to several members of TNF receptor superfamily, such as *TNFRSF1A, TNFRSF10A, TNFRSF10B*, and FAS receptors and are involved in the regulation of a wide spectrum of biological processes having tumor regression among them (Aggarwal 2003;Fulda 2009). The appearance of the extrinsic pathway might have represented an adaptation of an ancient system that has been developed for ontogeny control (i.e. apoptosis *per se*) of new functions.

Complex multicellular organisms composed of both post-mitotic and renewable tissues are cancer prone. Differently, simple multicellular organisms that are composed entirely or largely of post-mitotic somatic cells rarely develop cancer. As an example, the organisms *Drosophila melanogaster* and *Caenorhabditis elegans* do not develop cancer (Campisi, 2008). Accordingly, the importance of tumor suppressor mechanisms increase after vertebrata origin. This evolutionary stage matches the appearance of the extrinsic pathway and *TP53* suggesting an increment in tumor suppressor role of apoptosis. The *p53* protein not only regulates apoptosis, but is also a key regulator of cellular senescence. Cellular senescence is a tumor suppressor mechanism that leads to cellular morphological modifications and permanent cell cycle arrest, preventing damaged cells from dividing (Campisi, 2003; Zuckerman *et al.*, 2009). If senescence evolved jointly with *p53*, this mechanism might have evolved specifically to cancer suppression and it certainly represents an important field to understand cancer evolution.

References

Adimoolam, S. and Ford, J. M. (2002). p53 and DNA damage-inducible expression of thexeroderma pigmentosum group C gene. *PNAS*, 99, 12985-12990.

Aggarwal B. B. (2003). Signalling pathways of the TNF superfamily: a double-edged sword. *Nat. Rev. Immunol.*, 3:745-756.

Alano, C. C.; Ying, W. and Swanson, R. A. (2004). Poly(ADP-ribose) Polymerase-1-mediated Cell Death in Astrocytes Requires NAD+ Depletion and Mitochondrial Permeability Transition. *J. Biol. Chem.*, 279, 18895-18902.

Albertson, D. G.; Collins, C.; McCormick, F. and Gray, J. W. (2003). Chromosome aberrations in solid tumors. *Nature genetics*,34, 369-376.

Ameisen, J. C. (2002). On the origin, evolution, and nature of programmed cell death: a timeline of four billion years. *Cell Death Differ.*, 9, 367-393.

Aoki-Kinoshita K. F. and Kanehisa M. (2007). Gene annotation and pathway mapping in KEGG. *Methods Mol. Biol.*, 396, 71-91.

Atlas de Mortalidade por Câncer. Instituto Nacional de Câncer (INCA). Available at: <http://mortalidade.inca.gov.br/>Accessed 4 February 2009.

Backvall, H.; Asplund, A.; Gustafsson, A.; Sivertsson, A.; Lundeberg, J.; and Ponten, F. (2005). Genetic tumor archeology: microdissection and genetic heterogeneity in squamous and basal cell carcinoma. *Mutation Research/Fundamental and Molecular Mechanisms of Mutagenesis*, 571, 65-79.

Baker, P. G.; Goble, C. A.; Bechhofer, S.; Paton, N. W.; Stevens, R. and Brass, A. (1999). An ontology for bioinformatics applications. *Bioinformatics*, 15, 510-520.

Barabasi, A. L. and Oltvai, Z. N. (2004). Network biology: Understanding the cell's functional organization. *Nature Reviews Genetics,*5, 101-113.

Barrett, T.; Troup, D. B.; Wilhite, S. E.; Ledoux, P.; Rudnev, D.; Evangelista, C.; Kim, I. F.; Soboleva, A.; Tomashevsky, M.; Marshall, K. A.; Phillippy, K. H.; Sherman, P. M.; Muertter, R. N. and Edgar R. (2009). NCBI GEO: archive for high-throughput functional genomic data. *Nucleic Acids Res.*, 37: 885–890.

Benjamini, Y. and Hochberg, Y. (1995). Controlling the False Discovery Rate - A Practical and Powerful Approach to Multiple Testing. *Journal of the Royal Statistical Society Series B-Methodological*, 57, 289-300.

Berczi, C.; Bocsi, J.; Bartha, I.; Math, J. and Balazs, G. (2002). Prognostic value of DNA ploidy status in patients with rectal cancer. *Anticancer Research,* 22, 3737-3741.

Bild, A. H.; Yao, G.; Chang, J. T.; Wang, Q.; Potti, A.; Chasse, D.; Joshi, M.B.; Harpole, D.; Lancaster, J. M.; Berchuck, A. *et al.* (2006). Oncogenic pathway signatures in human cancers as a guide to targeted therapies. *Nature,*439, 353-357.

Bittner, T., *et al.* Individuals, universals, collections: on the foundational relations of ontology. In: FormalOntology in Information Systems (2004) Amsterdam, The Netherlands: IOS Press. 37–48. *Proceedings of the 3rd International Conference – FOIS*, 2004.

Bradley, N. (2002).*The XML Companion* (*3rd* edition). London: Addison-Wesley.

Bult, C. J.; Eppig, J. T.; Kadin, J. A.; Richardson, J. E. and Blake, J. A. (2008). The Mouse Genome Database (MGD): mouse biology and model systems. *Nucleic Acids Res.*, 36, D724-D728.

Campisi, J. (2003). Cancer and ageing: rival demons? *Nat. Rev. Cancer*, 3, 339-349.

Campisi, J. (2008). Aging and cancer cell biology. *Aging. Cell*, 7, 281-284.

Carroll, S. B. (2001). Chance and necessity: the evolution of morphological complexity and diversity. *Nature*, 409, 1102-1109.

Castro, M. A. A.; Filho, J. L.R.; Dalmolin, R. J. S.; Sinigaglia, M.; Moreira, J. C. F.; Mombach, J. C. M. and de Almeida, R. M. C. (2009). ViaComplex: software for landscape analysis of gene expression networks in genomic context. *Bioinformatics*, 25, 1468-1469.

Castro, M. A. A.; Mombach, J. C. M.; de Almeida, R. M. C.; Moreira, J. C. F. (2007). Impaired expression of NER gene network in sporadic solid tumors. *Nucl. Acids Res.*, 35, 1859-1867.

Castro, M. A. A.; Onsten T. G. H.; Moreira, J. C. F. and de Almeida, R. M. C. (2006). Chromosome aberrations in solid tumors have a stochastic nature. *Mutation Research/Fundamental and Molecular Mechanisms of Mutagenesis*, 600, 150-164.

Castro, M. A. A.; Onsten, T. T. G.; de Almeida, R. M. C. and Moreira, J. C. F. (2005). Profiling cytogenetic diversity with entropy-based karyotypic analysis. *Journal of Theoretical Biology,* 234, 487-495.

Castro, M. A.; Dalmolin, R. J.; Moreira, J. C.; Mombach, J. C. and de Almeida, R. M. (2008). Evolutionary origins of human apoptosis and genome-stability gene networks. *Nucleic Acids Res.*, 36, 6269-6283.

Chen, H. Y.; Yu, S. L.; Chen, C. H.; Chang, G. C.; Chen, C. Y.; Yuan, A.; Cheng, C. L.; Wang, C. H.; Terng, H. J.; Kao, S. F. et al. (2007). A Five-Gene Signature and Clinical

Outcome in Non-Small-Cell Lung Cancer. *The New England Journal of Medicine*, 356, 11-20.

Christie, K. R.; Weng, S.; Balakrishnan, R.; Costanzo, M. C.; Dolinski, K.; Dwight, S. S.; Engel, S. R.; Feierbach, B.; Fisk, D. G.; Hirschman, J. E. et al. (2004). Saccharomyces Genome Database (SGD) provides tools to identify and analyze sequences from Saccharomyces cerevisiae and related sequences from other organisms. *Nucleic Acids Res.*, 32, D311-D314.

Cleaver, J. E; Thompson, L. H.; Richardson, A. S. and States, J. C. (1999). A summary of mutations in the UV-sensitive disorders: xeroderma pigmentosum, Cockayne syndrome, and trichothiodystrophy. *Hum. Mutat.*, 14, 9-22.

Crespi, B. and Summers, K. (2005).Evolutionary biology of cancer. *Trends in Ecology and Evolution*, 20, 545-552.

Danial, N. N. and Korsmeyer, S. J. (2004). Cell Death: Critical Control Points. *Cell*, 116, 205-219.

Duesberg, P.; Li, R.; Fabarius, A. and Hehlmann, R. (2005). The chromosomal basis of cancer. *Cell Oncol.*, 27, 293-318.

Feinberg, A. P.; Ohlsson, R. and Henikoff, S. (2006). The epigenetic progenitor origin of human cancer. *Nature Reviews Genetics*, 7, 21-33.

Flyger, H. L.; Larsen, J. K.; Nielsen, H. J. and Christensen, I. J. (1999). DNA ploidy in colorectal cancer, heterogeneity within and between tumors and relation to survival. *Cytometry*, 38, 293-300.

Fulda S. (2009). Caspase-8 in cancer biology and therapy. *Cancer Lett.*, 281, 128-133.

Futreal, P. A.; Coin. L.; Marshall. M.; Down, T.; Hubbard, T.; Wooster, R.; Rahman, N. and Stratton, M. R. (2004). A Census of Human Cancer Genes. *Nature Reviews Cancer*,4, 177-183.

Gatenby, R. A. and Frieden, B. R. (2004). *Mutation Research/Fundamental and Molecular Mechanisms of Mutagenesis,* 568, 259.

Glockner, S.; Buurman, H.; Kleeberger, W.; Lehmann, U. and Kreipe, H. (2002). Marked intratumoral heterogeneity of c-myc and CyclinD1 but not of c-erbB2 amplification in breast cancer. *Laboratory Investigation*, 82,1419-1426.

Greenman, C.; Stephens, P.; Smith, R.; Dalgliesh, G. L.; Hunter, C.; Bignell, G.; Davies, H.; Teague, J.; Butler, A.; Stevens, C. *et al.* (2007). Patterns of somatic mutation in human cancer genomes. *Nature*, 446, 153-158.

Guarino, N. (1998) Formal Ontology in Information Systems. Amsterdam, The Netherlands: IOS Press. 3–15.*Proceedings of FOIS'98*, Trento, Italy, June 6–8.

Harada, K.; Nishizaki, T.; Ozaki, S.; Kubota, H.; Ito, H. and Sasaki, K. (1998). Intratumoral cytogenetic heterogeneity detected by comparative genomic hybridization and laser scanning cytometry in human gliomas. *Cancer Research*, 58, 4694-4700.

Helling, R. B. (1968). Selection of a Mutant of Escherichia coli Which Has High Mutation Rates. *J. Bacteriol.*, 96, 975-980.

Heng, H. H.; Liu, G.; Bremer, S.; Ye, K. J.; Stevens, J. and Ye, C. J. (2006). Clonal and non-clonal chromosome aberrations and genome variation and aberration. *Genome*, 49, 195-204.

Hipfner, D. R. and Cohen, S. M. (2004). Connecting proliferation and apoptosis in development and disease. *Nature Reviews Molecular Cell Biology*, 5, 805-815.

Hoeijmakers, J. H. J. (2001). Genome maintenance mechanisms for preventing cancer. *Nature*, 411, 366-374.

Hoglund, M.; Frigyesi, A.; Sall, T.; Gisselsson, D.; and Mitelman, F. (2005). Statistical behavior of complex cancer karyotypes. *Genes Chromosomes and Cancer*, 42, 327-341.

Hoglund, M.; Gisselsson, D.; Sall, T. and Mitelman, F. (2002). Coping with complexity. multivariate analysis of tumor karyotypes. *Cancer Genetics and Cytogenetics,*135, 103-109.

Hooper, S. D. and Bork, P. (2005). Medusa: a simple tool for interaction graph analysis. *Bioinformatics*, 21, 4432-4433.

Huettenbrenner, S. ; Maier, S. ; Leisser, C. ; Polgar, D. ; Strasser, S. ; Grusch, M. and Krupitza, G. (2003). The evolution of cell death programs as prerequisites of multicellularity. *Mutation Research/Reviews in Mutation Research*, 543, 235-249.

Hwang, B. J.; Ford, J. M.; Hanawalt, P. C. and Chu, G. (1999). Expression of the *p48* xeroderma pigmentosum gene is p53-dependent and is involved in global genomic repair. *PNAS*, 96, 424-428.

Jass, J. R.; Kelloff, G. J. and Schilsky, R. L. (2004). Limitations of the Adenoma-Carcinoma Sequence in Colorectum. *Clinical Cancer Research*, 10, 5969-5970.

Jefford, C. E. and Irminger-Finger, I. (2006). Mechanisms of chromosome instability in cancers. *Critical Reviews in Oncology/Hematology,*59, 1-14.

Jensen, L. J.; Kuhn, M.; Stark, M.; Chaffron, S.; Creevey, C.; Muller, J.; Doerks, T.; Julien, P.; Roth, A.; Simonovic, M. et al. (2009). STRING 8--a global view on proteins and their functional interactions in 630 organisms. *Nucl. Acids Res.*, 37, D412-D416.

Jeong, H.; Mason, S. P.; Barabasi, A. L. and Oltvai, Z. N. (2001).Lethality and centrality in protein networks. *Nature,* 411, 41-42.

Jin, C.; Jin, Y.; Wennerberg, J.; Akervall, J.; Dictor, M. and Mertens, F. (2002). Karyotypic heterogeneity and clonal evolution in squamous cell carcinomas of the head and neck. *Cancer Genetics and Cytogenetics*, 132, 85-96.

Jiricny, J. (2006). The multifaceted mismatch-repair system.*Nature Reviews Molecular Cell Biology*, 7, 335- 346.

Jones, S.; Chen, W.D.; Parmigiani, G.; Diehl, F.; Beerenwinkel, N.; Antal, T.; Traulsen, A.; Nowak, M. A.; Siegel, C.; Velculescu, V. E. *et al.* (2008). Comparative lesion sequencing provides insights into tumor evolution. *Proceedings of the National Academy of Sciences,*105, 4283-4288.

Kanehisa, M.; Goto, S.; Hattori, M.; Aoki-Kinoshita, F. Itoh, *M.*; Kawashima, S.; Katayama, T.; Araki, *M. and*Hirakawa, M. (2006). From genomics to chemical genomics: new developments in KEGG.*Nucl. Acids Res.*, 34, D354-D357.

Kendal, W. S. (1990).The use of information theory to analyze genomic changes in neoplasia.*Mathematical biosciences*, 100, 143-159.

Kulasingam, V. and Diamandis, E. P. (2008). Strategies for discovering novel cancer biomarkers through utilization of emerging technologies. *Nat. Clin. Prac. Oncol.*,5, 588-599.

Lash A. E. et al. (2000). SAGEmap: a public gene expression resource.*Genome Res.* 10, 1051-1060.

Lengauer, C.; Kinzler, K. W. and Vogelstein, B. (1998). Genetic instabilities in human cancers. *Nature,*396, 643-649.

Librelotto, G. R.; Ramalho, J. C.; Henriques, P. R.; Gassen, J. B. and Turchetti, R. C. (2008).A framework to specify, extract and manage Topic Maps driven by ontology. In SIGDOC. ACM, 2008.

Line, M. A. (2002). The enigma of the origin of life and its timing. *Microbiology*, 148, 21-27.

Loeb, L. A.; Loeb, K. R. and Anderson, J. P. (2003). Multiple mutations and cancer.*PNAS,* 100, 776-781.

Losi, L.; Baisse, B.; Bouzourene, H. and Benhattar, J. (2005). Evolution of intratumoral genetic heterogeneity during colorectal cancer progression. *Carcinogenesis*, 26, 916-922.

Ludwig, J. A. and Weinstein, J. N. (2005). Biomarkers in Cancer Staging, Prognosis and Treatment Selection. *Nature Reviews Cancer*, 5, 845-856.

Maley, C. C.; Galipeau, P. C.; Finley, J. C.; Wongsurawat, V. J.; Li, X.; Sanchez, C. A.; Paulson, T. G.; Blount, P. L.; Risques, R. A. ; Rabinovitch, P. S., *et al.* (2006). Genetic clonal diversity predicts progression to esophageal adenocarcinoma. *Nat. Genet.*, 38, 468-473.

Marx, J. (2002). *Debate Surges Over the Origins of Genomic Defects in Cancer.Science,* 297, 544-546.

Meier, P.; Finch, A. and Evan, G. (2000). Apoptosis in development. *Nature*, 407, 796-801.

Merlo, L. M. F.; Pepper, J. W.; Reid, B. J. and Maley, C. C. (2006). Cancer as an evolutionary and ecological process. *Nature Reviews Cancer*, 6, 924-935.

Mombach, J. C.; Castro, M. A.; Moreira, J. C. and de Almeida, R. M. (2008). On the absence of mutations in nucleotide excision repair genes in sporadic solid tumors. *Genet. Mol. Res.*, 7, 152-160.

Ogino, S. and Goel, A. (2008). Molecular Classification and Correlates in Colorectal Cancer. *Journal of Molecular Diagnostics*, 10, 13-27.

Osterheld, M. C.; Laurini, R.; Saraga, E. and Bosman, F. T. (1999). Evaluation of heterogeneity of DNA ploidy in early gastric cancers. *Anal. Cell Pathol.*,19, 67-72.

Park, J. and Hunting, S. (2003). XML Topic Maps: *Creating and Using Topic Maps for the Web*, volume ISBN 0-201-74960-2. Addison Wesley.

Potti, A.; Mukherjee, S.; Petersen, R.; Dressman, H. K.; Bild, A.; Koontz, J.; Kratzke, R.; Watson, M. A.; Kelley, M.; Ginsburg, G. S. *et al.* (2006). A Genomic Strategy to Refine Prognosis in Early-Stage Non-Small-Cell Lung Cancer. *The New England Journal of Medicine,*355, 570-580.

Rajagopalan, H.; Nowak, M. A.; Vogelstein, B. and Lengauer, C. (2003). *Nature reviews.Cancer,* 3, 695.

Ransohoff, D. F. (2005). Bias as a threat to the validity of cancer molecular-marker research. *Nature Reviews Cancer*, 5, 142-149.

Ravasz, E.; Somera, A. L.; Mongru, D. A.; Oltvai, Z. N. and Barabási, A. L. (2002). Hierarchical organizationof modularity in metabolic networks. *Science*, 297, 1551-1555.

Remm, M.; Storm, C. E. V. and Sonnhammer, E. L. L. (2001). Automatic clustering of orthologs and in-paralogs from pairwise species comparisons. *Journal of Molecular Biology*, 314, 1041-1052.

Rubbi, C. P. and Milner, J. (2003). *EMBO Journal*, 22, 975.

Sengupta, S. and Harris, C. C. (2005). p53: Traffic cop at the crossroads of DNA repair and recombination. *Nature Reviews Molecular Cell Biology*, 6, 44-55.

Setlow, R. B. and Carrier, W. L. (1964). Disappearance of Thymine Dimers from Dna - Error-Correcting Mechanism. *Proceedings of the National Academy of Sciences of the United States of America*, 51, 226-231.

Shannon, C. E. (1948). A mathematical theory of communication. *Bell System Technical Journal,* 27, 379-423.

Shen, L., Toyota, M., Kondo, Y., Lin, E., Zhang, L., Guo, Y., Hernandez, N.S., Chen, X., Ahmed, S., Konishi, K., *et al.* (2007). Integrated genetic and epigenetic analysis identifies three different subclasses of colon cancer. Proceedings of the National Academy of Sciences, 104, 18654-18659.

Shibata, D. (2006). Clonal diversity in tumor progression. *Nat. Genet.*, 38, 402-403.

Sjoblom T. et al. (2006). The Consensus Coding Sequences of Human Breast and Colorectal Cancers. *Science*, 314 ,268-274.

Snel, B.; Lehmann, G., Bork, P., Huynen MA. (2000). STRING: a web-server to retrieve and display the repeatedly occurring neighbourhood of a gene. *NucleicAcids Research*, 28, 3442-3444.

Venkatesan, R. N.; Bielas, J. H. and Loeb, L. A. (2006). Generation of mutator mutants during carcinogenesis. DNA Repair 5:294-302.

Vogelstein, B. and Kinzler, K. W. (2004). Cancer genes and the pathways they control.*Nature Medicine*, 10, 789-799.

von Mering C. Jensen, L. J.; Snel, B.; Hooper, S. D.; Krupp, M.; Foglierini, M.; Jouffre, N.; Huynen, M. A. and Bork, P. (2005). STRING: known and predicted protein–protein associations, integrated and transferred across organisms.*Nucl.Acids Res*. 33, D433-D437.

Wache, H.; Vögele, T.; Visser, U.; Stuckenschmidt, H.; Schuster, G.; Neumann, H. and Hübner, S. (2001). Ontology-based integration of information— a survey of existing approaches. In H. Stuckenschmidt, editor, IJCAI–01 Workshop: Ontologies and Information Sharing, 108–117.

Wang, Z.H., Cummins, J.M., Shen, D., *et al.*Three classes of genes mutated in colorectal cancers with chromosomal instability. *Cancer Res.*, 64, 2998-3001.

Wheeler D. L.; Church,D. M.; Federhen, S.; et al.(2003). Database resources of the National Center for Biotechnology. *Nucl.Acids Res*. 31, 28-33.

Whitfield, J. (2004).Origins of life: born in a watery commune. *Nature*, 427, 674-676.

Wildenberg, J. and Meselson, M. (1975). Mismatch Repair in Heteroduplex DNA. *PNAS*, 72, 2202-2206.

Willetts, N. S. and Clark, A. J. (1969).Characteristics of Some Multiply Recombination-Deficient Strains of Escherichia coli. *J. Bacteriol.* , 100, 231-239.

Wood, L. D.; Parsons, D. W.; Jones, S.; Lin, J.; Sjoblom, T.; Leary, R. J.; Shen, D.; Boca, S. M.; Barber, T.; Ptak, J. *et al.* (2007). The Genomic Landscapes of Human Breast and Colorectal Cancers. *Science*, 318, 1108-1113.

Wood, R. D.; Mitchell, M. and Lindahl, T. (2005). Human DNA repair genes, 2005.*Mutation Research-Fundamental and Molecular Mechanisms of Mutagenesis,* 577, 275-283.

Yan, B.; Wang, H.; Peng, Y.; Hu, Y.; Wang, H.; Zhang, X.; Chen, Q.; Bedford, J.S.; Dewhirst, M. W. and Li, C. Y. (2006). A unique role of the DNA fragmentation factor in maintaining genomic stability. *PNAS*, 103, 1504-1509.

Zhivotovsky, B. and Kroemer, G. (2004). Apoptosis and genomic instability. *Nat. Rev. Mol. Cell Biol.*, 5, 752-762

Zuckerman, V.; Wolyniec, K.; Sionov, R. V.; Haupt, S. and Haupt, Y. (2009). Tumour suppression by p53: the importance of apoptosis and cellular senescence. *J. Pathol.*, 219, 3-15

In: Bioinformatics Research: New Developments
Editors: C. Battik and K. Belhassine
ISBN: 978-1-61942-363-3

Chapter IX

Protein Bioinformatics for Drug Discovery Concavity Druggability and Antibody Druggability

Hiroki Shirai[1,*], Kenji Mizuguchi[2], Daisuke Kuroda[3,4], Haruki Nakamura[3], Shinji Soga[1], Masato Kobori[1] and Noriaki Hirayama[5]

[1] Advanced Genomics, Molecular Medicine Research Laboratories, Astellas Pharma Inc., Japan
[2] National Institute of Biomedical Innovation, Saito, Ibaraki-city, Osaka, Japan
[3] Institute for Protein Research, Osaka University, 1-3 Yamadaoka, Suita-city, Osaka, Japan
[4] Graduate School of Frontier Biosciences, Osaka University, Suita-city, Osaka, Japan
[5] Tokai University School of Medicine, Isehara, Japan

Abstract

The field of protein bioinformatics analyzes the sequence and structure of protein; it plays a critical role in the discovery of small therapeutic agents as well as protein drugs. Here we present our recent progress in this field, including the new concepts of concavity druggability and antibody druggability, which are expected to raise the drug discovery success ratio.

Keywords: Concavity druggability, antibody druggability, protein bio-informatics, drug discovery

Introduction

Currently, the two *in silico* technologies, bioinformatics and computational chemistry, are effectively used for drug discovery[1-6].

Bioinformatics plays various roles in drug discovery at multiple stages: target discovery, target validation, building a screening system, and analysis of the mechanism of diseases as well as that of compounds[1-3]. It also contributes to the generation of protein drugs and small therapeutic agents. It comprises four components: information technology (IT), information science, theoretical biology, and theoretical biochemistry, although the boundaries between them are obscure. The IT aspect of bioinformatics includes the construction and maintenance of an information infrastructure (hardware, software, network systems, and in-house database). Various kinds of biological data are collected and entered into the in-house database to yield a user-friendly interface and easily accessible data. The information science aspect of bioinformatics is the analysis of microarray and other biological data using sophisticated data mining approaches. Text mining, which extracts useful information from literature, falls into this category. The theoretical biology aspect of bioinformatics, sometimes called pathway bioinformatics, is used to determine which genes are involved in physiological functions. The theoretical biochemistry aspect of bioinformatics, which is almost the same as protein bioinformatics, includes the analysis of protein sequence or structure in order to elucidate its molecular function. In addition, protein bioinformatics naturally plays a critical role in protein drug discovery.

Despite the importance of protein bioinformatics, inefficient information-sharing between industrial and academic researchers in this field could prevent the development of more useful tools for real drug discovery. The surrent status of protein bioinformatics and problems therewith are summarized and addressed below.

Just ten years ago, there were still significant difficulties with generating a lead compound via docking studies between compounds and a target protein. The improvement of software and hardware as well as the rapid increase in protein/compound interaction data over the past decade have enabled us to obtain lead compounds in many projects. In addition to docking studies, other technologies involving computational chemistry effectively contribute to drug discovery at various stages[4-6]. The structure-activity relationship is used for lead optimization, ADME and toxicity prediction for lead evaluation, and chemoinformatics for clustering and the rational selection of compounds. More importantly, the generation of the "druggability of small compounds" concept and its suitability as an indicator for lead compound evaluation changed the general flow of drug discovery.

However, despite efforts made based on computational chemistry, compounds are often dropped for various reasons. Researchers try to reduce the risk of this happening by evaluating ADME and toxicity at an earlier stage, increasing the variety of lead compounds, and optimizing as many of them as possible. However, it should be noted that the use of computational chemistry alone would reduce only some specific risks and generate a limited variety of lead compounds. A combination approach, such as with bioinformatics, is more desirable as it is expected to reduce different types of risks while increasing lead compound variety.

Bioinformatics and computational chemistry are currently used independently for drug discovery; however, if used in combination, a synergistic effect is expected. Protein bioinformatics is considered to be a way to combine these two *in silico* technologies; thus, it plays a key role in the creation of an integrated solution, which may improve the next generation drug discovery success ratio. Effective usage of sequence and structure analysis could reduce side effects. This matter is described in detail in the next section. In addition, we recently proposed a new concept, concavity druggability, which is expected to increase the variety of lead compounds (described in section II).

I. Protein Sequence and Structure Analysis

The efficiency of protein bioinformatics in drug discovery is illustrated below by introducing our work with the guanidino-group modifying enzyme (GME) superfamily[7-9].

The GME superfamily consists of enzymes (GMEs) that catalyze the modification of guanidino groups[7]. This superfamily includes many key metabolic enzymes, some of which are already recognized as attractive drug targets. GMEs adopt a unique tertiary structure known as the α/β propeller, which can accommodate a diverse set of sequences. The amino-acid sequence identities among GMEs are from 8% to 23%. Amidinotransferase (AT) is the first enzyme of this superfamily to have its structure determined. It catalyzes the transfer of amidino group from amino acid arginine to another substrate. The crystal structure of AT revealed a barrel-shaped fold with a cavity on one side where the substrate arginine binds. The chemical reaction is catalyzed by a Cys-His-Asp catalytic triad at the bottom of this cavity.

Our first project dealing with the protein bioinformatics of GME was to propose the existence of the GME superfamily itself[8]. Before our proposal, despite the functional similarity, the low sequence similarities among the enzymes and the lack of the structural information except for AT prevented the researchers from understanding their evolutionarily relatedness (homology) for a long time. FUGUE is the sophisticated fold recognition software which one of the authors (Kenji Mizuguchi) developed [10]. By using FUGUE, we predicted three other enzyme families to be homologous to AT and share similar catalytic mechanisms [8]. These are dimethylarginine dimethylaminohydrolase (DDAH), Arginine deiminase (ADI) and Porphyromonasgingivalis peptidyl-arginine deiminase (PPAD) and related hypothetical proteins showing weak similarity to it (which we called PPADH). After our prediction, the structures of DDAH, ADI and PPADH were determined by X-ray crystallography, and confirmed our hypothesis[11-13].

The second GME project was the prediction of succinyl arginine dihydrolase (AstB) as a new member of the GME superfamily[9]. After prediction, the structure of AstB was determined by X-ray crystallography, which confirmed our hypothesis[14].

The useful information obtained from these two drug discovery projects are summarized as follows: i) Structures useful in docking studies for determining compounds that block GME can be predicted. ii) Catalytic residues of DDAH, ADI, PPAD, PPADH, and AstB; this information is valuable for validating protein as a drug target as well as useful for setting up an assay system. iii) Identification of PPADH proteins as new drug targets. They would have some enzymatic activity in common with GME, and produce ammonia. Proteins that generate ammonia sometimes act as virulence factors in pathogenic microorganisms, thus the PPADH from *Helicobacter pylori* might act as such by generating ammonia from the abundant peptides in the stomach. iv) Catalytic mechanisms of DDAH, ADI, PPAD and AstB. These catalytic mechanisms are valuable for the design of chemical compounds at the lead optimization step.

The sequence analysis aspect of protein bioinformatics could extract these pieces of valuable information prior to structure determination by X-ray crystallography.

Mammalian peptidyl-arginine deiminase (PAD) is a metallo-enzyme that had been considered unrelated to GME until the crystal structure of human PAD was determined[15]. The structure of PAD is composed of several domains, one of which is the α/β propeller. The overall structural similarity, the conservation of catalytic residue, and the similarity of function confirmed that it is another member of the GME superfamily. Its relatedness could not be predicted prior to structural determination; therefore it is necessary to improve the sensitivity of the fold recognition tools.

After many GME structures had been determined, we conducted a third GME project[7]. Structural superposition and structure-based alignment could identify new key GME residues involved in catalysis and substrate binding. We found that conserved guanidino-carboxyl interactions are used in two different ways: acidic residues in the catalytic site form hydrogen bonds to the substrate guanidino group, and the enzyme Arg residues at several key positions recognize the carboxyl group of the substrate and fix its orientation. Based on this observation, we proposed rules for classifying the GME sequences and predicting their molecular function from the conservation of the key acidic and Arg residues.

The useful predictive tools generated by this third drug discovery project are summarized below. i) The GME commonality and diversity information extracted allows for the design of a compound with high selectivity. ii) The proposed rules allow for the prediction of pathogenic organism unannotated gene function and the proposal of new drug targets.

Importantly, protein bioinformatics can be used to extract valuable information even after structural determination by X-ray crystallography.

Generally, if a compound binds to its target protein, it can also bind to the homologues of the target, which would eventually generate a side effect. For this reason, we usually screen not only for the target protein, but also counter-screen for its homologues to select compounds with both high affinity and selectivity. Selecting the homologues to be checked is crucial, but there is some controversy over whether it is necessary to check remote homologues or not. Since remote homologues are so different, some researchers insist that only close homologues need to be checked, while others insist that all should be. We feel that checking depends on the features of the superfamily to which the target protein belongs. For example, the GME superfamily comprises enzymes capable of a diverse array of reactions (transferase, hydrolase, and dihydrolase), and whose substrates are the same or similar to each other (arginine or related ligands). Despite the very low sequence similarity among GMEs, they are considered to bind to the same or similar compounds. Thus, if finding a compound that blocks a certain GME with high selectivity is desired, the selectivity of remote homologues should also be checked. In contrast, the acyl-CoA N-acyltransferase (NAT) superfamily comprises enzymes whose substrates are diverse, but whose reactions are the same (acyl transferase)[9]. For this reason, NATs are not considered to easily bind to the same or similar compounds. Thus, if a compound that blocks a certain NAT with high selectivity is desired, checking the selectivity of remote homologues is not likely to be necessary. Using protein bioinformatics is expected to reduce the number of different types of side effects that must be tackled with computational chemistry.

II. Concavity Druggability

Using protein bioinformatics, we proposed a new concept, "concavity druggability", and developed a method to evaluate druggability from the amino-acids present in the concavity on the protein surface[16,17].

The specific binding of a ligand to its target protein is the key to drug action. The surface of a target protein usually possesses multiple concavities where small molecules may bind. However, each ligand binds preferentially to a specific concavity. We defined druggable concavity as the concave surface where drug-like molecules are highly inclined to bind. Drug targets that can be modified by a small molecular drug need to be identified for each disease phenotype. Thus, finding a druggable concavity in disease-related proteins is a crucial step towards validating new drug targets. It is also important for increasing the chance of obtaining a drug-like compound and for understanding the function of a given protein.

Since the binding site of a drug is considered to be highly specific to its molecular characteristics, the binding concavity must have a distinct character significantly different from those of similar concavities on protein targets. One of the authors (Noriaki Hirayama) recently established a profile for determining the "drug-likeness" of a compound, which comprises multiple molecular descriptors to determine how much like a drug the molecule is[18]. We evaluated the drug-likeness of ligands, which was determined while their structures were complexed with proteins from the Protein Data Bank (PDB). The amino acid compositions around the binding sites for drug-like ligands in well-qualified X-ray structures were analyzed in detail. The analysis revealed a remarkable propensity for the presence of specific amino acids at the binding site of each drug-like compound. From this data, a simple discrimination index called Propensity for Ligand Binding (PLB) was developed, which allows the druggability of concavities on a given protein surface to be evaluated. Importantly, the PLB index can be used to identify druggable concavities in homology models.

Use of the PLB index to find new druggable concavities would increase the variety of lead compounds that would be obtained if computational chemistry alone was used.

III. Antibody Modeling and Druggability

Protein bioinformatics contributes to the generation of protein drugs as well as small compound drugs. Recently, more antibody drugs are expected to be developed because of their low toxicity and high efficiency. Establishing high affinity binding is crucial for expanding detection limits, extending dissociation half times, decreasing drug dosage, and increasing drug efficacy[19]. Affinity maturation of antibodies *in vivo* often fails to generate antibody drugs of the targeted potency, which makes it necessary to perform further affinity maturation *in vitro* using directed evolution or rational design via protein bioinformatics.

For rational design, a tertiary structural model of an antigen binding site needs to be constructed from its amino acid sequence. The efficiency of *in silico* design depends on the accuracy of the model. An antigen-binding site is composed of six complementarity-determining regions (CDRs). Five of these CDRs (CDR-L1, L2, L3, H1, and H2) have a limited number of canonical structures; however, the third CDR of the heavy chain (CDR-H3) shows substantial diversity in length, sequence, and structure. In addition, CDR-H3 sometimes changes its conformation depending on the existence of an antigen. Importantly, it lies in the centre of the antigen-binding site and generally plays a dominant role in antigen recognition. Thus, building a model of CDR-H3 is the most important and the most difficult step.

After examination of the qualified crystal structures of the antibody, we proposed empirical rules for predicting the structural features of CDR-H3 from its amino-acid sequence (H3-rules)[20,21]. Essentially the same proposal was also made by a different group[22]. We recently revised the rules (H3-Rules 2007) to apply only to the analysis of newly determined structures[23]. Multicanonical molecular dynamics simulation is one of the algorithms for enhanced conformational sampling[24]. We found it is quite useful for building an accurate model and capturing the structural variety of CDR-H3s[25,26]. At present, the usage of revised H3-rules and multicanonical molecular dynamics simulation provides the best approach for determining the antigen binding site. However, it remains difficult to predict the structure of long CDR-H3s, the non-canonical structures of the other five CDRs, and the dimerization angle of the H chain and L chains. Thus, the methods of antibody model building need to be improved.

Current drug discovery practice uses the druggability of small compounds as an indicator of lead compound potential. However, a similarly rational approach has not yet been established for antibody evaluation. However, such rational approaches have not yet been established for antibody evaluation. We recently proposed a new concept, "antibody druggability"[23]. We examined the CDR-H3 structures of 12 antibody drugs and found the common structural features. Antibody druggability would be effectively applied to rational antibody design and selection.

Conclusion

Although protein bioinformatics plays a critical role in drug discovery, the inefficiency of information-sharing between industrial and academic researchers in this field could prevent the development of more useful tools for real drug discovery. In this review, our recent progress in protein bioinformatics was presented, and the current status of this field, along with associated problems that may be experienced by pharmaceutical companies, were summarized. It is our hope that more useful tools will be developed in this field to increase the success ratio of drug discovery even further.

References

[1] Chen YP, and Chen F. Identifying targets for drug discovery using bioinformatics. (2008) *Expert Opin Ther Targets* 12:383-389.

[2] Mizuguchi K. (2004) Fold recognition for drug discovery. *Drug Discovery Today: Targets;*3:18-23.

[3] Yan Q. (2008) The integration of personalized and systems medicine: bioinformatics support for pharmacogenomics and drug discovery. *Methods Mol Biol.*;448:1-19.

[4] Mohan CG, Gandhi T, Garg D, and Shinde R. (2007) Computer-assisted methods in chemical toxicity prediction. 1: *Mini Rev Med Chem.* 7:499-507.

[5] Tropsha A, and Golbraikh A. (2007) Predictive QSAR modeling workflow, model applicability domains, and virtual screening. *Curr Pharm Des.*;13:3494-3504.

[6] Taft CA, Da Silva VB, Da Silva CH. (2008) Current topics in computer-aided drug design. *J Pharm Sci.* 97:1089-1098.

[7] Shirai H, Mokrab Y, and Mizuguchi K. (2006) The guanidino-group modifying enzymes: structural basis for their diversity and commonality. *Proteins.* 64:1010-1023.

[8] Shirai H, Blundell TL, and Mizuguchi K. (2001) A novel superfamily of enzymes that catalyze the modification of guanidino groups. *Trends Biochem Sci.* 26:465-468.

[9] Shirai H, and Mizuguchi K. (2003) Prediction of the structure and function of AstA and AstB, the first two enzymes of the arginine succinyltransferase pathway of arginine catabolism. *FEBS Lett.* 555: 505-510.

[10] Shi J, Blundell TL, and Mizuguchi K. (2001) FUGUE: sequence-structure homology recognition using environment-specific substitution tables and structure-dependent gap penalties. *J Mol Biol.* 310: 243-257.

[11] Murray-Rust J, Leiper J, McAlister M, Phelan J, Tilley S, Santa-Maria J, Vallance P, and McDonald N. (2001) Structural insights into the hydrolysis of cellular nitric oxide synthase inhibitors by dimethylarginine dimethylaminohydrolase. *Nat Struct Biol.* 8: 679-683.

[12] Das K, Butler GH, Kwiatkowski V, Clark AD Jr, Yadav P, and Arnold E. (2004) Crystal structures of arginine deiminase with covalent reaction intermediates; implications for catalytic mechanism. *Structure* 12: 657-667.

[13] Galkin A, Kulakova L, Sarikaya E, Lim K, Howard A, and Herzberg O. (2004) Structural insight into arginine degradation by arginine deiminase, an antibacterial and parasite drug target. *J Biol Chem.* 279: 14001-14008.

[14] Tocilj A, Schrag JD, Li Y, Schneider BL, Reitzer L, Matte A, and Cygler M. (2005) Crystal structure of N-succinylarginine dihydrolase, AstB, bound to substrate and product, an enzyme from the arginine catabolic pathway of Escherichia coli. *J Biol Chem.* 280: 15800-15808.

[15] Arita K, Hashimoto H, Shimizu T, Nakashima K, Yamada M, and Sato M. (2004) Structural basis for Ca(2+)-induced activation of human PAD4. *Nat. Struct. Mol. Biol.* 11: 777-783.

[16] Soga S, Shirai H, Kobori M, and Hirayama N. (2007) Use of amino acid composition to predict ligand-binding sites. *J Chem Inf Model.* 47:400-406.

[17] Soga S, Shirai H, Kobori M, and Hirayama N. (2007) Identification of the druggable concavity in homology models using the PLB index. *J Chem Inf Model.* 47:2287-2292.

[18] Horio K, Muta H, Goto J, and Hirayama N. (2007) A simple method to improve the odds in finding 'lead-like' compounds from chemical libraries. *Chem. Pharm. Bull.* 55: 980-984.

[19] Lippow SM, Wittrup KD, and Tidor B. (2007) Computational design of antibody-affinity improvement beyond in vivo maturation. *Nat. Biotechnol.* 10: 1171-1176.

[20] Shirai H, Kidera A, and Nakamura H. (1996) Structural classification of CDR-H3 in antibodies. *FEBS Lett.* 399: 1-8.

[21] Shirai H, Kidera A, and Nakamura H. (1999) H3-rules: identification of CDR-H3 structures in antibodies. *FEBS Lett.* 455: 188-197.

[22] Morea V, Tramontano A, Rustici M, Chothia C, and Lesk AM. (1998) Conformations of the third hypervariable region in the VH domain of immunoglobulins. *J Mol Biol.* 275: 269-294.
[23] Kuroda D, Shirai H, Kobori M, and Nakamura H. (2008) Structural classification of CDR-H3 revisited: A lesson in antibody modeling. *Proteins* inpress.
[24] N. Nakajima, H. Nakamura and A. Kidera (1997) Multicanonical ensemble generated by molecular dynamics simulation for enhanced conformational sampling of peptides. *J. Phys. Chem* B101: 817–824
[25] Shirai H, Nakajima N, Higo J, Kidera A, and Nakamura H. (1998) Conformational sampling of CDR-H3 in antibodies by multicanonical molecular dynamics simulation. *J Mol Biol.* 278: 481-496.
[26] Kim ST, Shirai H, Nakajima N, Higo J, and Nakamura H. (1999) Enhanced conformational diversity search of CDR-H3 in antibodies: role of the first CDR-H3 residue. *Proteins*. 37: 683-696.

Index

E

F

J

K

L

M

Q

R

S

T

U

V

W

X

Y

Z